A Genetic and Cultural Odyssey

Linda Stone and Paul F. Lurquin

A Genetic and Cultural Odyssey

The Life and Work of **L. Luca Cavalli-Sforza**

COLUMBIA UNIVERSITY PRESS / NEW YORK

Columbia University Press
Publishers Since 1893
New York Chichester, West Sussex

Library of Congress Cataloging-in-Publication Data
Stone, Linda, 1947–
A genetic and cultural odyssey : the life and work of L. Luca Cavalli-Sforza / Linda Stone and Paul F. Lurquin.
p. cm.
Includes bibliographical references and index.
ISBN 0-231-13396-0 (cloth : alk. paper)
1. Cavalli-Sforza, L. L. (Luigi Luca), 1922– 2. Geneticists—California—Biography I. Lurquin, Paul F. II. Title.

QH429.2.C28S76 2005
576.5'092—dc22
[B]

2004061869

Columbia University Press books are printed on permanent and durable acid-free paper.

Printed in the United States of America

c 10 9 8 7 6 5 4 3 2 1

To the memories of Alfredo Trombetti and Joseph Greenberg, linguists, Alfred Kroeber, anthropologist, and Ronald Fisher, geneticist

Contents

Preface

This book was written by an anthropologist (Linda Stone) and a geneticist (Paul Lurquin). Our subject is L. Luca Cavalli-Sforza, a Stanford University professor whose work has combined our two fields. When we first met Cavalli-Sforza in 1994, we had no idea that we would one day write his biography. He had come to speak at Washington State University, where we are both professors, and we had talked with him briefly at a reception following one of his talks. It was a few years later that we began to discuss the career of this polymath individual who uniquely bridges the humanities and the experimental sciences. These discussions convinced us that a scientific biography of Cavalli-Sforza's life and work should be written. In this way we stumbled upon our own interdisciplinary collaboration, which we hoped would be in keeping with the open and collaborative spirit of Cavalli-Sforza's own work.

What impressed us about Cavalli-Sforza's career was not just its scope—he has moved from his signal discoveries in bacterial genetics to his contributions to the study of human evolution and culture—or its depth—he has made theoretical contributions to both human physical and cultural evolution. What struck us most was the way in which Cavalli-Sforza's work intersects science with society, a venture attempted by very few. He has used genetics to help solve questions of human origins and development, and in the process his work has stirred up a number of social issues, some of them quite controversial. In turn, social issues have impacted Cavalli-Sforza's work; as with all scientists, his work has taken place within and has been shaped by the times in which he lived. This book, an authorized biography, tells the story of Cavalli-Sforza's career, tracing these connections between science and society.

In 2002, with the production of his biography now seriously in mind, we contacted Cavalli-Sforza and discussed our project with him. With his support for our idea, we then arranged for sabbatical leaves from our university to devote full-time work to this project.

The resulting book is based on two primary activities. First, we read Cavalli-Sforza's major books and articles—a daunting task given the large number of his publications—as well as related works by others. Second, we conducted a series of in-depth interviews with Cavalli-Sforza. We also carried out briefer interviews with some of his colleagues. For this we made frequent visits to Palo Alto over 2002–2004. All our interviews were tape recorded and later transcribed. In the end we had accumulated more than twenty hours of taped interviews with Cavalli-Sforza.

We interviewed Cavalli-Sforza mostly in his office at the Stanford University School of Medicine but also sometimes in his home in Palo Alto. We also met him for lunch and dinner on a few occasions, once also with his son, Francesco. In addition, we spent a whole weekend with him in Portland, Oregon, on his stopover between Seattle and Italy. We also exchanged dozens of e-mail messages with him over the course of almost two years.

Cavalli-Sforza could not have been a more gracious and cooperative host for our study. He was always willing to meet with us and devote whatever time we needed to provide us with details and commentary on his life and work. Indeed his stamina during interviews surpassed our own; often when we got tired and wanted to call it a day, he was ready and willing to continue. Thanks to him, our stays in Palo Alto were very pleasant experiences and our research went smoothly.

This book is intended for the serious reader. It is a scientific biography that covers in some depth the fields of both genetics and anthropology as they relate to Cavalli-Sforza's contributions. It is written for students, professionals, or others interested in human evolution, physical and cultural, and how that subject has been studied over the last half of the twentieth century. It tells a story of where human curiosity, if given enough freedom, can lead. In this book, we strive to put Cavalli's work in a historical and scientific context. However, our intention never was to write a review of all the fields that his work has touched. Innumerable contributions have been made by others to the fields of bacterial and human population genetics, cultural anthropology, archaeology and linguistics. A comprehensive exposition of all these topics would be impossible within the scope of a biography. Nevertheless, we draw attention to areas of Cavalli's research that have raised (and are still raising) questions in the scientific community.

The title of our book contains the word "odyssey." This is intentional. Instead, we could have used the word "adventure," which basically means the same thing, and this is indeed the term we originally planned to use. However, while discussing the title with us, Cavalli was reminded of his high school days during which, like every other Italian student, he studied Dante's *Divine Comedy*. In canto XXVI of "Inferno," Dante meets Ulysses (Odysseus) in Hell.

There, he hears from Ulysses, who is no longer the cunning trickster of the Trojan War, the tale of his last adventure, which goes as follows: After finally returning to his home in Ithaca, Ulysses and his companions set sail again, this time to the west. As they are crossing the Strait of Gibraltar, Ulysses harangues his companions in this way: "Brothers, who through a hundred thousand dangers have come with me to the extreme west, do not deny to the little part of life that still remains to us, the experience of discovering the land without people that lies behind the sun. You were not made to live like brutes, but to follow valor and knowledge" (translated by L. L. Cavalli-Sforza).

While recognizing the grandiloquence of Ulysses' declaration, Cavalli likes it because it conveys a spirit of adventure coupled with a search for knowledge. This, we believe, characterizes very well Cavalli's trips to the four corners of the earth to uncover our origins.

We are well aware that many statements made in this book will be contested by various geneticists and anthropologists. This is not only because our biographical subject is a controversial figure, but also because both human evolutionary genetics and anthropology, and particularly research involving their combinations, are currently controversial and under debate in many respects. Since many of the topics covered in this book are so contested, any account of them will inevitably show biases. Our bias, we admit, is in favor of the work and ideas of our subject, Cavalli-Sforza.

We are deeply grateful to Professor Cavalli-Sforza for welcoming us in his house and in his Stanford office. We also thank his close collaborators at Stanford—Marcus Feldman, Merritt Ruhlen, Peter Oefner, and Peter Underhill—for kindly accepting to be interviewed by us. Our colleague at Washington State University, Barry Hewlett, who also works closely with Cavalli-Sforza, is thanked for introducing us to him, facilitating our visits to Stanford, and exchanging his views on the genetics-anthropology combination with us. We wish to thank John Bodley for his comments on particular chapters of the book, Robert Ackerman for discussing Beringia (see glossary) with us, and our research assistant, Matthew Landt, for his help with literature searches. We also much appreciate the help of Phyllis Mayberg, Cavalli-Sforza's secretary, in our efforts to procure copies of many of the numerous articles he has written. We also thank Luca Cavalli-Sforza, Marcus Feldman, Peter Underhill, Merritt Ruhlen, as well as six anonymous reviewers, for reading all or portions of our manuscript and/or prospectus. We are deeply grateful to Sir Walter Bodmer for his meticulous critique of our manuscript and allowing us to interview him at Stanford. We warmly thank Joshua Lederberg for answering our questions regarding his collaboration with Cavalli-Sforza in the early days of bacterial genetics. Finally, we are grateful to Robin Smith, senior editor for the sciences at Columbia University Press, as well as our copyeditor, Roy Thomas, for their consistently helpful assistance with the manuscript.

Cavalli-Sforza also kindly provided many of the photographs presented in this book. To help the flow of the book's writing, we have avoided citations in the text. However, an extensive list of references, on a chapter by chapter basis, appears at the end of the book. As always, all remaining errors are ours.

This project was supported in part through a sabbatical leave granted to the authors by Washington State University and by a grant from the Honors College of Washington State University.

A Genetic and Cultural Odyssey

Chapter 1

Science and Society, Genes and Culture

One Saturday afternoon in 1988 a linguist, Merritt Ruhlen, walked into the JJ&F grocery store in Palo Alto, California. To his surprise he saw there a colleague, a Stanford University scientist, a tall, distinguished-looking Italian man, already prominent in his field of genetics. When they met in the store the scientist excitedly told the linguist about the positive first results of their collaboration: the mapping of human population gene frequencies on the globe was showing a strong correlation with the mapping of human language families! The scientist was so excited he grabbed a napkin and, no doubt amid strange glances from other grocery shoppers, drew out how the languages and gene frequencies correlated.

This scientist is Luigi Luca Cavalli-Sforza (fig. 1.1), and these first correlations between genes and languages as etched on the napkin were soon to become a part of his important contributions to the unraveling of human prehistory. Our purpose is to tell his story and the development of his fecund career, which spans over sixty years. That period saw World War II; recovery in Europe; major advancements in science, especially genetics; student revolutions on both sides of the Atlantic; and the growth of an antiscience sentiment in some academic disciplines. Our book covers the ways that these events and others were interwoven with the career of L. L. Cavalli-Sforza.

At the time of writing, Professor Cavalli-Sforza (hereafter, Cavalli) is eighty-three years old. He is a very active professor emeritus at Stanford University's School of Medicine. He rides his bike to work, continues to travel all over the world, drives a spiffy red Mercedes, and enjoys a glass of wine with dinner. With over five hundred scientific articles and ten books, his work has integrated a number of scientific fields to address questions of human origins and diversification, often inspiring new insights, sometimes inspiring controversial social issues. In this process, his work has been at the interface of science and society.

Figure 1.1 Portrait of Professor L. Luca Cavalli-Sforza (1999). (From the collection of L. L. Cavalli-Sforza)

In this book, we hope to demonstrate that multidisciplinary research is still possible in our age of hyperspecialization and reductionism. We also want to show that controversy can be the gist of science, and that bold attempts at unifying seemingly disparate fields can lead to interesting thoughts and models of the origins of contemporary human beings. One of our aims is equally to show that a scientific career can take mysterious paths—for example, leading from an investigation of bacterial sex to the study of human origins, as done by Cavalli in his long career.

The two fields most closely integrated in Cavalli's career are genetics and anthropology. Although very different in many respects, both deal with similar concepts. Both are interested in descent in the form of kinship and cultural transmission for the anthropologist, and descent through gene transmission for the geneticist. Kinship, cultural and gene transmission, evidently, all rely on the ability of living organisms to engage in sexual reproduction. While the mechanisms of sexual activity differ widely among species, the ultimate result is invariably the same: offspring contain the indelible prints of their ancestors.

Before elaborating how genetics and anthropology each approach transmission (covering the genetic and anthropological concepts used throughout this book) and the consequences of this, we will first discuss the interface of scientific and societal issues.

Science and Society

It is undeniable that scientific discoveries have had, have now, and will have in the future an impact—sometimes profoundly, sometimes less so—on society. We will use two examples to illustrate this science-society interface—the cases of Darwin and Einstein, two scientists whose names are recognized by nearly everyone. Both men are either praised or criticized by some segments of society, even though they themselves could not have foreseen how much their basic, nontechnological research would change the world.

Charles Darwin (1809—1882), more than 120 years after his death, still deeply polarizes society at one of its fundamental levels, that of religion. Surveys regularly show that over 40 percent of the American public believes that Earth and humans were created about six thousand years ago, in a process that took exactly six days. This is the case in spite of the overwhelming scientific evidence to the contrary and Darwin's great theory that upset this scenario.

As we know, Darwin developed the idea of descent by modification and theorized evolution by **natural selection.*** Darwin's revolutionary idea, based on an enormous compendium of observational data, was to imagine that forces of nature such as climate, volcanic eruptions, abundance or lack of food, predators, and other factors, literally selected—determined—which individuals in a group would survive and proliferate, and which would not. This natural selection was made possible by subtle differences that existed between various individuals, differences that made them suitable (or not) to survive and reproduce under changing natural circumstances. Given enough time, individual differences could lead to the appearance of new species and the disappearance of others. It is important to emphasize that evolution does not entail the notion of progress, or betterment. A fish is not a step above seaweed, except in complexity. With evolution, life forms do become more complex and sophisticated, but they do not necessarily represent advances over earlier forms, at least in terms of reproductive success.

Nevertheless, many members of the public do see evolution as an upward progress, with humans at the top of what the Aristotelians used to call the "Great Chain of Being." In our view, this is a dangerous way of looking at the grand picture of evolution because it can easily lead to the assertion that certain human populations are superior to other human groups.

In any event, Darwin's theory of evolution suggested that different living species appeared at different times, and that younger ones evolved from older

---

* Throughout this book, bold type indicates the first relevant textual use of a term that can be found in the glossary.

ones through a gradual process. Many older species, like the dinosaurs, became extinct and new species emerged. In brief, the biblical story of the special and separate creation of species had been toppled.

Darwin's theory of course met with great hostility among the clergy and most of the rest of society. This was to be expected, because a materialistic explanation of the living world was now threatening to supplant an age-old, Christian religious dogma. What is more, the biblical myth, as well as the moral justification for the existence of church hierarchies and social classes, were in danger of being replaced by rational thinking, no longer religious faith. In addition, Darwin's theory of evolution was consistent with the new sciences of geology and paleontology, most of whose practitioners could no longer conceive of Earth as being created in 4,004 B.C.E. (on Sunday, October 23, at 9:00 A.M. to be precise) , as claimed by Bishop James Ussher in 1658.

Darwin's detractors resorted to personal attacks to discredit him, including apparently asking Thomas Huxley, his ardent supporter, whether he descended from apes through his grandmother or through his grandfather (this comment may be apocryphal since what was said at the meeting between Huxley and Bishop Wilberforce, Darwin's detractor, was not recorded). This shows these people's poor understanding of the theory of evolution, which already stated in its original form that *humans do not descend from apes as they exist today.* They simply share with apes a common ancestor which existed millions of years ago and is now extinct. Today, the theory of evolution is accepted by all biologists, excepting for the very small—but vocal—minority of those who declare themselves "creationists" or proponents of "intelligent design."

Yet in spite of a mountain of experimental evidence, including the observation of evolutionary forces at work in the laboratory and in nature, the understanding and acceptance of evolution are dismal among the American public. This is best illustrated by the constant attempts, often successful, to present a "balanced" view of the origins of life and the cosmos in the classroom. This balanced view simply means teaching the Judeo-Christian myths of creation in parallel with the scientific evidence that supports evolution. This is of course not the place to enter this debate. We simply want to underscore that, in the case of Darwinism, science has met society and, for better or for worse, has polarized it. An excellent proof of this is the escalation in the number and variety of bumper stickers visible in the United States. A number of years ago, devout Christians affixed to their cars the symbol of a fish, standing for and showing the old Greek acronym ΙΧΤΥΣ (*fish* in Greek), meaning Jesus Christ Savior Son of God. Soon thereafter, noncreationists started showing the same fish, but this time with the name DARWIN inside. Interestingly, the fish had sprouted legs; it was evolving. The latest installment in this bumper-sticker war (so far) is a small befooted DARWIN fish being swallowed by a much bigger—and legless—fish called TRUTH. One can be sure that this is not the end of the controversy begun by Charles Darwin, and it is not impossible to imagine a reiteration of the famous (or perhaps infamous) Scopes trial.[1]

As for Albert Einstein (1879–1955), he is rightly famous for his two theories of relativity, the special and the general. Special relativity tells us that the speed of light cannot be exceeded, that time and mass vary with velocity, and that mass and energy are basically two different facets of the same reality. In other words, energy can, under proper circumstances, be converted into massive objects, and the latter can be converted into energy. General relativity, on the other hand, is a theory of space and gravitation. It tells us that gravity results from the warping of space by massive objects, and it explains very well the properties of the cosmos, from the expansion of the universe to the existence of black holes. In spite of having radically modified the way we think about the universe as a whole, it cannot be said that general relativity has had much of an impact on society at large. Not so for special relativity, however.

Hiroshima, McCarthyism, backyard fallout shelters, school drills, the Cold War, the potentially massive escalation of the Indo-Pakistani conflict, the notion of "rogue" nations, the threat of nuclear terrorism as well as the threat (but also the potential) of nuclear power stations (the latter being not necessarily always a bad thing), all squarely originate from one single discovery: the conversion of mass (in the form of uranium or plutonium) into energy (in the A-bomb or in the nuclear reactor's core). In other words, Einstein's 1905 groundbreaking and at the time seemingly innocent discovery turned the world into a more dangerous place. Of course, Einstein should not be blamed for any of this. He was a dedicated pacifist and regarded with horror the military applications of his theoretical work. Nonetheless, his is a clear example of a profound and far-reaching meeting, perhaps ultimately a clash, between science and society.

---

1. The Scopes trial is also known as the "Monkey trial." In 1925, John T. Scopes, a high school biology teacher in Dayton, Tennessee, was sued for teaching evolution in the classroom, a practice then banned by the state of Tennessee. Scopes was eventually found guilty by a jury and forced to pay a fine of $100. However, the forceful defense by his attorney Clarence Darrow, and his brilliant demonstration that William Jennings Bryan, attorney for the state of Tennessee, was in fact advocating the violation of the constitutional barrier between church and state, swayed public opinion in the favor of Scopes. One year later, the Tennessee Supreme Court reversed the Dayton jury decision. Elsewhere in the world, the Scopes trial was ridiculed and used as a criticism of scientifically backward, fundamentalist America. Today, several state legislatures are advancing the same backward attitude. This time, however, these people advocate the teaching of creationism (or so-called intelligent design, creationism's more intellectually dressed-up but equally bogus cousin) to high school students for the sake of "fairness." This is tantamount to saying that the validity of science can be decided at the ballot box.

These two examples illustrate one of the points we want to develop in this book—that science and society, or culture, can never be dissociated. For better or for worse, the former influences the latter and the latter the former. We will see also in this book that some social scientists believe that the exalted and hegemonic position of science in Western society is misplaced. In this view, an academic current sometimes referred to as **postmodernism**, science itself is a mere cultural construction. Taken to the extreme, this position sees science as, like all cultural constructions, dependent on human subjectivities, with no real, objective value.

In the social sciences, postmodernism has raised questions about academic or disciplinary authority. For example, in cultural anthropology, postmodernists question on what authority an anthropologist, usually an outsider in the culture he or she studies, could claim to validly represent or interpret that culture. How could the anthropologist's statements about that culture possibly be considered more true or more valid than the visions of themselves and their own society held by the people in that culture? Is not the anthropologist merely interpreting the other culture in terms of his or her own culturally biased concepts and categories? In the larger picture, whose vision of the world gets to be the correct one? By extension, on what basis should any so-called "scientific" vision of the world be considered superior to any other vision—for example, a vision of the world that centers on beliefs in ghosts and witches, gods and demons, or what have you? Science in this view is not metacultural, on a plane above any cultural vision. It is a way of thinking developed in the Western world, ultimately no more true, real, or valid than any other way of thinking.

As we will explore later, postmodernism is important in understanding some of the controversies that have grown around Cavalli's work. Here, one key issue is "identity." Cavalli's work has given a genetic definition to human groups, with considerable commentary on their past, their ancestry, their migrations, their prehistory and history. What happens when this definition and commentary conflict with the ideas and traditions of these peoples themselves, concerning who they are, who their ancestors were, and where they came from? What happens when an individual's or a group's self-proclaimed identity is tied to land rights and other property but is unsupported genetically? Authority over human identity—who gets to say "who's who?"—has become a major social issue around the globe.

As with all scientists, Cavalli's work has taken place within a particular cultural context as well as a particular time in history. The fact that Cavalli's field is genetics and that a great deal of his work has been conducted in the United States is significant given the particular cultural meanings and perceptions of genetics within America. American culture expresses both a fascination with and a fear of genetics and genetic knowledge. The fascination is easy to see in American popular culture. For example, hardly an episode goes by on daytime soap operas where the term "DNA" is not mentioned at least once. Soap characters frequently request DNA tests to determine whether another character really is one's parent, sibling, or child; further spicing up the plots, these characters often deceive one another by somehow faking the tests. Yet a fear of genetic knowledge combined with new technologies is also culturally expressed, for example in debates over the morality of surrogate motherhood and, more recently, human cloning. It was unsettling enough in the late twentieth century to see the splitting up of "motherhood," to realize that the genetic mother of a child is not necessarily also the birth mother. Now with human therapeutic cloning apparently achieved and human reproductive cloning on the horizon, we wonder even more where genetic knowledge and new technologies will take us, or how this new science will force us to redefine human identity and human family relationships. The fact that the first claims of human cloning were orchestrated by the Raelians—a new religious sect claiming that human DNA was first brought to earth by space aliens—makes us question the security of even the scientific boundaries of genetics.

A kind of love-hate relationship with genetics is culturally expressed in America in other ways as well. Medically speaking, genetic counseling is considered good and useful for couples planning a child; and the discovery of "disease genes" can help medical professionals treat patients or can help patients understand their susceptibility to certain diseases. At the same time, many people in the United States and elsewhere are very concerned about the harmful effects of genetically modified organisms to the ecology and to the very food we buy and consume. Science fiction novels and films like *The Boys from Brazil* and *Jurassic Park* play upon these fears of genetic tampering and its potentially devastating consequences.

We can also see in American culture an ambivalent attitude toward the power of genetics over the question of human identity, reflecting currents of postmodernism, as discussed above. On the one hand, there is a traditional view that genetic knowledge is definitive of who we are and how we are related to others. In the United States, one's "real" father, for example, was for a long time understood to be exclusively the biological father, whose DNA one shares, no matter how caring a step-, foster, or adoptive father might be. Recently this cultural vision has been challenged by another that seeks to define a person's closest family relationships through individual choice and will, downplaying the significance of genetic sharing and transmission. This new vision has undoubtedly come about as the number of American stepparents, stepchildren, and step-siblings has grown with rising divorce and remarriage, and as novel family forms—for example, gay and lesbian married couples with children—have also emerged. This discussion demonstrates that the public is no longer neutral or indifferent toward genetics, one of Cavalli's chosen fields.

Before moving to Cavalli's work and, in subsequent chapters, showing the connections between his work and social issues, we provide here a brief introduction to the two fields Cavalli combines, genetics and anthropology. Readers already familiar with basic concepts in genetics and anthropology may want to skip the next several sections and go directly to chapter 2.

How Geneticists Think

Genetics is the science of inheritance. In most cases, inheritance is the consequence of organisms being able to exchange genes through sexual reproduction. This mechanism was probably "invented" well over three billion years ago. Back then, sexual matings probably had occurred already among very simple organisms since higher life forms had not yet appeared. Slowly, these organisms evolved into more complex cells that themselves gave rise to plants, animals, and more recently, humans. During these billions of years of evolution, sexual reproduction, as a mechanism that allows genetic diversification, was never forgotten by nature. However, as we shall see, sexual reproduction is only one of several mechanisms that lead to biological diversification.

One of the goals of geneticists is to understand how and when the many millions of living species that inhabit our planet evolved from one another and how this great genetic diversity came to be. The study of the origin of genetic diversity and relatedness between organisms is called **phylogeny**, from the Greek words *phylon* (class, category) and *genesis* (origin, generation). Science views the diversification of complex life forms as starting from a common ancestor, a population of unicellular organisms that lived about 1.6 billion years ago. This common ancestor—sometimes called the last common ancestor—evolved from simpler bacterial cells that first appeared about 3.5 billion years ago. The last common ancestor then differentiated into plants and animals through a branching process. Worms appeared about 1 billion years ago, followed by insects (600 million years ago), fish (400 million years ago), reptiles and birds (300 million years ago), mammals (200 million years ago, and the genus *Homo* (2 million years ago). The main point here is that all living creatures are linked to simpler ancestors that lived, evolved, and branched out in the past. In that way, all present life forms keep in their genes the "memory" of the creatures that preceded them and evolved into them. We will see in chapter 3 how simple statistical techniques are used to compute the genetic differences among organisms. But first, modern phylogenetic studies would not be possible without an understanding of what genes are, how they are passed on from parents to progeny, and how they can change.

The Nature of Inheritance

Sexual transmission of **genes** from parents to offspring was elucidated by Gregor Mendel in 1865. Mendel, who worked with plants, demonstrated that genes, the units of inheritance, behaved as "particulate" elements in sexual crosses. His theory further allowed prediction of the outcome of crosses involving several sharply contrasting characters. We now know that genes are responsible for the physical attributes of an organism, a set of properties that we call the **phenotype**. Phenotypic properties in plants include stem length, seed shape, and flower color, or other characteristics. In humans, phenotypes include height, eye color, blood group, premature baldness, and the ability to digest lactose, for example. Many phenotypes are not directly visible because they deal with molecules rather than readily apparent traits.

What is more, and this is very important for the rest of our story, Mendel understood that genes that determine a trait can exist as variations on a theme. Indeed, his experiments showed that a gene coding for flower color in peas, for example, can exist in the form of a "purple" gene or that of a "white gene." In other words, the generic gene that determines flower color in peas exists in the form of at least two variants, one giving flowers a purple color, the other one giving white. Human genes follow the same pattern. We now know, for example, that the generic gene that determines the ABO blood groups comes in three different variants, *A*, *B*, and *O*. Many human genes in a population of individuals exist in several different forms, as shown by the examples of eye and skin color, two traits that are genetically determined. However, modern genetic analysis goes well beyond these simple visible traits; geneticists can now directly analyze in fine detail the DNA of organisms, as well as the proteins that their DNA codes for.

The Nature of Genes and Their Function

Almost everyone knows that genes are made of **DNA**, short for deoxyribonucleic acid. Most also know that DNA is a double helical molecule whose structure was discovered in 1953 by James Watson, Francis Crick, Rosalind Franklin, and Maurice Wilkins. In genetic terms, the most important feature of DNA is the existence of base pairs in the center of the double helix. The base pairs form the "rungs" of the DNA "twisted ladder." It is the *sequence* of these base pairs in the DNA that determines the genes carried by organisms, from viruses to humans. All human beings have in their cells DNA molecules comprised of over 3 billion base pairs that contain about 25,000 genes. The suite of all these 3 billion base pairs is called the **genome**, and the slight base-pair variations that exist among the DNAs of different individuals determine their different **genotype**. The genotype of a person thus defines what **gene variants** exist within that person, and in turn, the phenotype of that person is determined by his/her genotype.

The precise functioning of DNA was unraveled in the two decades that followed the discovery of the double helix. Basically, DNA can be thought of as the blueprint of the cell, a molecular "instructions manual" of sorts. The information stored in the DNA base sequence is translated, through many complicated steps, in the form of **proteins**, the agents that are responsible for the phenotypes of cells and organisms. In other words, whether one's blood is A, B, or O, or whether a flower is purple or white, depends on the type of proteins made by the organism. These proteins can also be thought of as variants of a generic theme, much like genes can exist as variants of a generic DNA theme. Geneticists can easily determine who carries which type of variant simply by determining the base sequence of that individual's DNA or by studying his/her proteins. In short, which protein variant one harbors (one's phenotype) is entirely determined by which gene variant (one's genotype) one has.

The Origin of Gene Variants

Looking at a large collection of human beings, one sees that human phenotypes vary considerably. Much of this phenotypic variation is due to genotypic differences. When one compares the proteins and DNA sequences of these individuals, much more variation can be distinguished than by the naked eye because not all slight variations at the molecular level have visible consequences. Two major factors are responsible for this wide variation. First, we receive genes from our parents. This means that our DNA is a combination of two genomes. Whatever genetic variation was present in a mother and father may now be represented in their offspring. Even though the external phenotype of a particular child may not be much like that of the parents, DNA analysis can determine paternity and maternity with an extremely high degree of accuracy. Therefore, an individual can be seen as a sort of genetic sum of his or her ancestors. Some special segments of human DNA, those associated with **mitochondria** and the Y **chromosome**, are inherited from only one of the two parents, however (see chapter 6).

A second factor that explains genetic diversity is the spontaneous change of DNA base pairs in a person's genome. Sometimes such a change can be deleterious and lead to genetic disease, but often it can be completely "silent" and detectable at the DNA and protein levels only. Base-pair changes, as well as base-pair addition or removal, are known as **mutations**. To understand changes in DNA base pairs, one should think of these base pairs as ordinary chemicals that can interact in a variety of ways. DNA contains four different **bases**. Normally, the DNA base called adenine (A) forms a pair with another base called thymine (T). Conversely, the base called guanine (G) pairs with cytosine (C). Quite rarely, a wrong base pair can form, such as a G-T pair, an A-C pair, or even pairs between unusual bases present in DNA under chemical attack. The result of this is a change in the DNA base sequence of a gene and, consequently, a change in the structure of the protein this gene codes for. Changes like this are inherited and can be detected through DNA and protein studies. Since these changes are passed on to the progeny, the lineage of an individual can be traced to one or several ancestors in whom the DNA change took place. In summary, the study of proteins and DNA can be used to determine whether various individuals share common ancestors.

Today, there exist fast and reliable techniques for the study of proteins and DNA base-pair sequences. In fact, the full sequence of human DNA is known, and it is no longer difficult to analyze in great detail the many specific gene sequences that are known to vary in humans. Applications of these DNA and protein techniques by forensic geneticists are well known, but forensics is but one aspect of genetic technology.

How Geneticists Study the Past Using Protein and DNA Information

In addition to forensics, geneticists are interested in exactly how the thousands of genes carried by humans influence their health. Other geneticists are interested in biotechnology, and manipulate plant and animal genes for agricultural purposes. It may also be that, one day, biotechnological approaches will be used commonly to treat diseases. But that is not all: the study of genes and their products—proteins—also can be used to reconstruct the past.

Traditionally, reconstructing the past was done by paleontologists or physical anthropologists who study fossils and can date them with great accuracy using a variety of techniques. One might think that all geneticists would have to do is to isolate DNA from fossils and sequence it. This, with very rare exceptions, cannot be done because in most cases DNA does not survive in fossilized bone for more than a few thousand years. However, paleontologists have gathered much precious information regarding the phylogeny (the ancestry) of many species, including humans. By looking at fossils, it is possible to determine when in the past a particular species first appeared.

The theory of evolution shows that species appear from ancestral ones through a branching process assuming the general structure of a tree. These structures are called phylogenetic **trees**. Thus, the ancestry of modern humans can be traced back to a series of ancestral forms, thanks to the fossil record, all the way to tree shrews (fig. 1.2). Tree shrews, in turn, are derived from earlier mammals, themselves descending from reptiles, fish, and so on, with all organisms ultimately descending from an ancestor common to all plants and animals. Similarly, the fossil record has allowed us to determine when the closer ancestors of modern humans lived (fig. 1.3).

How, then, do geneticists use this fossil information to reconstruct the past, this time based on the DNA sequence of *organisms that are alive today*? The example given above can be used to illustrate the methodology in use. The fossil record shows that modern humans appeared more recently than Old World monkeys. The same record shows that Old World monkeys are more recent than tree shrews. Given these observations, and since tree shrews, Old World monkeys, and humans are still present today, it follows that tree

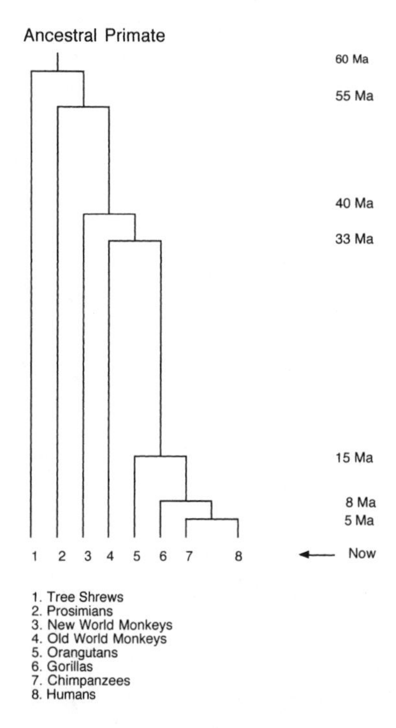

Figure 1.2 Phylogenetic tree of primate evolution. The ancestral primate appeared approximately 60 million years ago. Modern tree shrews are its direct descendants. Our closest relatives, chimpanzees, last shared a common ancestor with humans (protohumans, in fact) about 5 million years ago (Ma = million years ago). This date is generally agreed upon, although the human-chimpanzee split could have occurred as recently as 4.4 million years ago or as early as 7 to 8 million years ago. The first protohumans were *Australopithecus*, a bipedal species with a brain size approximately equal to that of a modern chimp.

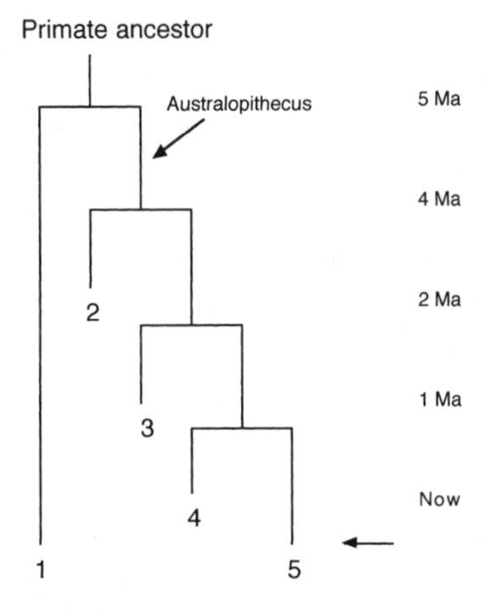

Figure 1.3 A highly simplified phylogenetic tree of hominids. The primate ancestor here is ancestral only to the African apes (gorillas, chimpanzees, and bonobos) and humans. This ancestor lived about 8 million years ago. *Australopithecus*, *Paranthropus*, *Homo erectus* (who invented fire), and *Homo neanderthalensis* are all extinct. The modern genus *Homo* has only one species, *sapiens*, ourselves.

shrews, when compared to both Old World monkeys and humans, *have had more time to accumulate differences in their DNA base pairs*. Thus, the number of differences found between Old World monkeys and humans, who both originated later in time, is expected to be less than the number of differences found between these two species and tree shrews, assuming a constant rate of mutation. In other words, when one compares modern, equivalent genes found in all three species, one expects that the base sequence of human genes should be closer to that of Old World monkeys than to that of tree shrews. This is what is in fact observed.

The procedure here is to use DNA sequence (or protein) diversification, over the time that has elapsed since first appearance, to determine when two species last shared a common ancestor. The longer the time, the more DNA base-pair changes in their genes. Thus, DNA from extant species can be used as a "molecular clock." The fossil record was (and is) used to calibrate that clock. From the fossil record, we know that chimpanzees and humans last shared a common ancestor about 5–7 million years ago. The sequencing of chimpanzee and human

DNAs shows a certain number of base-pair changes (differences) in specific genes. Knowing the number of these changes, therefore, we can calculate the *rate* of base-pair change per year and estimate the time-distance between the two species. Then, knowing this rate, we can isolate DNA from any living creature from any species, sequence it, look at the sequence for differences, compare this DNA to, say, human DNA as a standard, and calculate back the distance in time that separates that creature from us. We can do this because, at least within the simplified scope of this book, the molecular DNA clock "ticks" with regularity.

DNA and protein techniques can also be used to study the origin of modern humans, their diversification into ethnic groups, and their migrations across continents. Clearly, however, the time-distance that separates various human individuals in different populations is much less than the time-distance that separates humans from chimpanzees. Just as in the case of the chimpanzees, Old World monkeys, tree shrews, and humans, one can study the DNA sequence (or the proteins) of different human populations (or individuals) to determine at what times they diverged from ancestral *Homo sapiens*. Here, however, human phylogenies are mostly studied through the analysis of DNA and protein **polymorphisms**—that is, the types and numbers (or frequencies) of "variants" found in human populations.

A simple example helps clarify the concept of genetic divergence without recourse to polymorphisms, which will be explained in detail in later chapters. Let us assume for the sake of simplicity that chimps and humans diverged 5 million years ago. Let us also assume that a comparison between 100,000 base pairs of DNA present in a specific DNA segment reveals 500 differences between chimps and humans. In order to calculate the time-distance that separates Africans (the oldest human population in the world; see below) and non-Africans, for example, we need to know the average number of base-pair differences found in the same DNA region isolated from Africans and non-Africans. It turns out that the average number is 10. Thus, assuming a constant mutation rate, the average number of changes between Africans and non-Africans is 1/50th the number of changes found between humans and chimps. *Conclusion*: Africans diverged from non-Africans 50 times less time ago than chimps and humans diverged from one another. This period of time is 5 million years divided by 50—that is, 100,000 years, which corresponds well with archaeological data. This type of calculation was first used with a particular segment of the human genome called mitochondrial DNA in the laboratory of Allan Wilson at Berkeley in the late 1980s (see chapter 5).

The study of gene and protein polymorphisms among human populations is much more refined than the simple example given above and provides much more precise clues (see chapters 5 and 6) as to when and where humans diversified. A quick overview of the phylogeny of modern humans based on the study of gene and protein polymorphisms goes like this: Our common *H.*

sapiens ancestors appeared in Africa about 100,000 years ago. African people then differentiated first from these ancestors, followed by Australians and New Guineans (50,000 to 60,000 years ago), then followed by Asians and Caucasians. Last to differentiate were the Native Americans, somewhere between about 35,000 and 15,000 years ago.

A discussion of these studies, as well as anthropological studies that accompany them, constitutes the core of this book.

Culture and Anthropology

In the evolution of our species we came to rely less on biological changes and more on culture—that which can be learned and transmitted to others—as our primary means of adaptation. The human capacity for culture (a capacity shared with other animals but developed most in humans) was itself brought about through natural selection and in turn came to influence our physical evolution. It is probable, for example, that the development of human language was interwoven with the increase in human brain size and complexity. It is hard to specify exactly when this all began, and we certainly do not know exactly how or why. But it is here that the discipline of anthropology provides insights into human cultural adaptation and diversification.

Anthropology, as most readers will know, is the study of the physical, social, cultural, and linguistic development of human beings. Anthropologists are interested in describing particular cultures, past and present; studying how cultures develop and change over time; and in accounting for cross-cultural similarities and differences. A central tenet of most anthropology is **cultural relativism** (cultural relativity), or the idea that each culture should be understood and assessed in its own terms and not in the terms of any other culture. In this way, one culture cannot be held to be superior to another.

Anthropology is a broad discipline which in the United States conventionally divides into four subfields: physical anthropology, archaeology, cultural anthropology, and linguistics. Although anthropologists specialize in one of these areas, in anthropological research there is often considerable overlap between the four fields. We have already had a look at physical anthropology, which focuses on human evolution, nowadays using molecular genetic techniques, and the fossil record of this evolution, in the previous section.

Archaeology is the study of the human past through investigation of material remains of human societies. Archaeologists literally dig these remains out of the earth (or, in marine archaeology, under ocean water) in order to reconstruct cultures of the past. Archaeologists are interested in human **migrations** and the development and evolution of culture, both regionally and globally. The more glamorous archaeological discoveries—King Tut's tomb in Egypt, the terra-cotta soldiers of Xi'an, China—are only the tip of the iceberg of painstaking archaeological work that is as likely to involve counting pollen seeds in soil samples or brushing off endless chips of pottery as it will involve uncovering tombs and mummies.

Archaeology has developed a number of methods and techniques for locating good sites, excavating remains so that they are not destroyed and their context can be recorded, and for interpreting data. Central to archaeology are also methods for dating remains and establishing chronologies at excavation sites. One method, called stratigraphy, relies on the stratification of layers of earth in which remains are embedded. It yields relative dating: those objects found at the lower layers of the stratification are normally older than those found at higher layers. Dendrochronology, or tree-ring dating, can provide absolute dates. Useful only at particular sites where timber survives, this method makes use of the fact that trees add a ring each year during their growth. In particular ecological areas where a master sequence of tree rings with dates is available, pieces of excavated timber can be dated by year. Probably the most common method for absolute determinations of age is **carbon-14 dating.** This method relies on the fact that living organisms absorb radioactivity from the atmosphere. At the death of the organism, the remaining radioactivity decays at a known rate. Thus measuring the amount of radioactivity left in a piece of fossil insect, animal, plant, or human can provide an estimate of the amount of time that has elapsed since that organism was alive and so date the remains. Carbon-14 dating is restricted to organic remains and it is not reliable beyond about 40,000 years, or 60,000 with the expensive accelerator technique. Other radioactive elements are used to measure older dates.

Linguistics is the scientific study of human language. Its origins go back to ancient India when priests developed methods to analyze and so preserve Sanskrit, a sacred language used in Vedic rituals. Linguists study how the sounds of language can produce meaning. They study phonemes (distinct speech sounds in a given language) and their permissible combinations (phonology). Linguists also study morphology: how phonemes are combined to produce morphemes, or the minimal units of speech that have meaning. Thus, for example, the phonemes represented by the English letters *d*, *o*, and *g* combine to produce the morpheme *dog*. In reverse order they produce the morpheme *god*. The sounds alphabetically representable as "s," "zez," or "z" in English are morphemes that signal "plural" when attached to nouns, such as in, respectively, "hats," "roses," and "dogs." Linguists also look at language rules for combining morphemes into longer acceptable utterances, or sentences (syntax), and the rules by which these utterances become meaningful (semantics).

Linguists address a number of very interesting questions. When, how, and why did human language develop? What in the human brain makes language possible? Do chimpanzees (who can learn some sign language) have true language capability? How do human languages work? How do human children learn languages? How do specific human languages reflect particular cultural values and worldviews? Amid this wide range of questions, linguists also investigate contemporary social issues such as: Should elementary or secondary education be bilingual in various places? Many of these questions concern subfields such as sociolinguistics and psycholinguistics.

The field of linguistics also can be subdivided into descriptive linguistics—which looks at how language works—and historical linguistics—which studies how language changes and the historical relationships between languages. What is of particular relevance to Cavalli's work, and thus to our book, is historical linguistics. All languages change over time, and there are certain regularities in the process. Linguists can compare current languages, looking closely at their phonological, morphological, and syntactical similarities and variations, to determine which languages are historically related to one another. With this, linguists can classify languages into larger groupings of language families. For example, a number of languages (including English, Greek, Persian, French, Spanish, Hindi, and many others) are classified together in an Indo-European family of historically related languages that began to diverge from 6,000 to 10,000 years ago (the estimate varies). Based on comparisons of these languages, linguists can also at least partly reconstruct the ancient Indo-European **protolanguage**. There are many other families of languages (how many depends on which classification one follows and, outside of Indo-European languages, there is some disagreement among linguists on language relationships). How far back could this tracing of human language relationships go; or, could we trace all human protolanguages back to one ancestral human parent language? This is not known but, as we will see in later chapters, there is currently interest and research in this topic.

In addition to tracing relationships between languages, linguists can also estimate how long ago some languages diverged from one another, giving a time depth to historical linguistics. They use a technique called **glottochronology**. Glottochronology is based on the assumption (well founded from studies on languages with written records) that the percentage of words (called **cognates**, meaning words with a common origin) that two diverging languages share will decrease at a regular rate with time. This technique was developed by Morris Swadesh and Robert Lees, who showed how, with glottochronology, the time of divergence between two languages could be estimated on the basis of the percentage of cognates they now share. They also developed a standard word list (one hundred words that in languages generally tend to change least over time, such as words for objects in nature) to use in this kind of analysis. Unfortunately, glottochronology did not turn out to be as accurate or reliable as everyone had hoped, but it is still used for chronological estimates.

Cavalli himself, in collaboration with William Wang of Hong Kong University, has done work on the method of glottochronology. Using Micronesian languages, they showed that one problem with glottochronology is that dates of language separation are very approximate and tend to give underestimates for long time periods. Also, the rate of word replacement changes considerably between words. However, they were able to demonstrate that instead of looking at time, better results can be obtained by looking at the variation in word replacement in space (i.e., the measured distance between the current location of speakers of different but related languages), which can be measured more accurately than time of language separation.

Finally, cultural anthropology is the study of contemporary cultures. Cultural anthropologists study a wide range of topics, either within one culture or comparatively across cultures, such as social organization, kinship, religion and ritual, gender patterns, medical practices, economic life, structures of power and authority, and so on. A description of a particular culture along these and other lines is called an **ethnography**. Typically, cultural anthropologists spend a year or more living and conducting fieldwork in a particular society (often not their own) to produce an ethnography. Anthropologists use a number of different methods to study culture, but underlying nearly all of them is *participant observation*—the simultaneous process of directly participating in the life of a cultural group while recording observations of it. Within this participant observation, probably the most common research method for cultural anthropologists is the informal interview, simply asking local people many questions about their lives.

In terms of theory, cultural anthropology is and always has been in flux. Unlike the hard science of genetics (and also less like the subdisciplines of archaeology and physical anthropology), it does not appear to make cumulative advances in the collection of data or the understanding of its subject—culture—but rather reflects intellectual trends and fashions in the broader humanities. Today there are a number of different theoretical approaches in cultural anthropology. One currently popular approach emphasizes social process and human agency in relation to the structures of a society. This approach considers how individual choices and strategic actions can initiate social processes that either perpetuate or change societal structures. Many of those following this approach look at culture in terms of power structures and ways that these structures are perpetuated or resisted. A great deal of research on gender, class, and ethnicity falls into this theoretical frame. There are also materialist approaches (sometimes referred to as cultural materialism) which study culture in relation to material conditions of life, including ecological relationships, technology, and modes of economic production. Another, currently minor, theoretical area in cultural anthropology is evolutionary anthropology. Here the forces of cultural change are seen as both analogous to forces of human physical evolution and as interacting with them. It is in this area that Cavalli has made a theoretical contribution, as will be spelled out in chapter 4.

Whereas cultural anthropologists accept human biological evolution, the counterpart idea of human cultural evolution has met and continues to meet with resistance and discomfort. This reaction is in part due to the history of the concept of cultural evolution in the field. In the nineteenth century, a number of scholars, drawing analogies with Darwin's theory, posited that human cultures evolve in systematic ways through distinct stages from "lower" forms to "higher" forms. They often used pejorative words to indicate the lower stages. So, for example, Lewis Henry Morgan worked out an evolutionary scheme of cultural evolution proceeding from "savagery" through "barbarism" and finally to "civilization," based primarily on technological developments. The highest stage, "civilization," was based on the development of writing, but this stage also showed the development of monogamy (as opposed to the **polygyny** and promiscuity of earlier stages) and civil government. Morgan's scheme was but one of many such evolutionary schemes that saw modern Euro-American society as representing the highest level of cultural achievement and social progress. Though in time they too would evolve, contemporary hunting-gathering societies were seen as still stuck back in "savagery," and as representative of how all human societies used to be in the remote past.

This conception of cultural evolution was later discredited by anthropologists and accused of blatant **ethnocentrism** and racism. Many anthropologists considered that these evolutionary views violated what came to be the principle of cultural relativity. Making matters even worse, nineteenth-century evolutionary thought had also inspired the notion of social Darwinism, or the idea that human social life was a natural process of "survival of the fittest." Seeing human life as an inevitable competitive power struggle among individuals and groups, where only the fittest survive, social Darwinism justified social inequality. One social Darwinist, Herbert Spencer, advocated an end to charity to the poor (and the "backward races") because this aid went against the law of the survival of the fittest and only prolonged the misery of the poor and the "unfit."

As a result of this history, many cultural anthropologists continue to distance themselves from notions of cultural evolution. This is one reason why Cavalli's and others' revival of the concept of cultural evolution has not been met with enthusiasm by many anthropologists, even though Cavalli and other evolutionary scholars do not propose that some cultures are "more advanced than" or "superior" to others. As with biological evolution, the idea of cultural evolution need not and should not imply "progress," but a fear that it will persists in anthropology.

Whereas most anthropologists largely confine their research to either physical anthropology, archaeology, linguistics, or cultural anthropology, Cavalli has made use of anthropology in all four of its subfields, in combination with genetic data. What is significant about Cavalli's work is that he incorporates data from multiple sources, using, for example, linguistic data to cross-check data based on genetic variation. This has enabled him to serve as a kind of grand synthesizer in the study of human prehistory.

This biography traces Cavalli's career in a roughly chronological way. The next chapter discusses how Cavalli first became a medical doctor but later abandoned medicine to enter bacterial genetics, a field in which he made significant discoveries and contributions. Chapter 3 traces his move from bacterial genetics to human population genetics and recounts what major contributions Cavalli made to this field. Chapter 4 discusses Cavalli's theoretical work on cultural evolution as well as his theory of demic diffusion of Neolithic farmers who brought agriculture to Europe thousands of years ago.

Chapter 5 covers Cavalli's insightful joining of data from genetics, archaeology, and linguistics to investigate human prehistory and diversification. To give the reader a preview of the nature of Cavalli's work in the area of human diversification and migration, we show in figure 1.4 a grand summary of

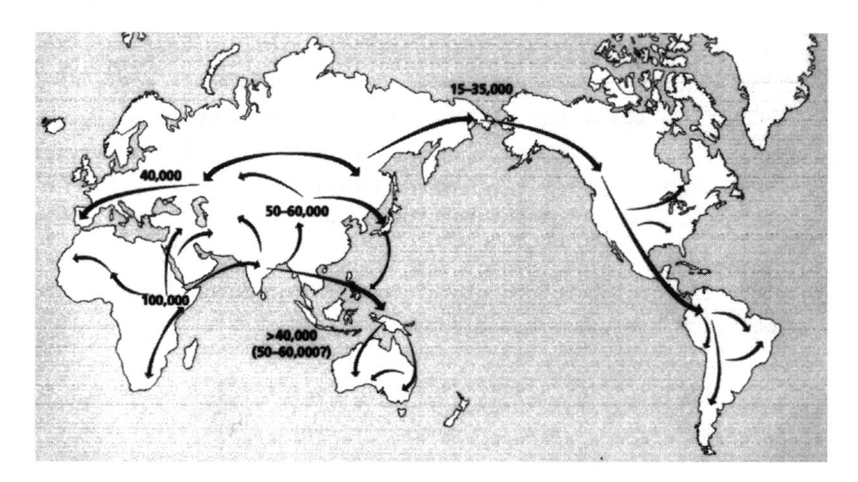

Figure 1.4 A map of the world indicating possible dates of departure and arrival of modern humans in various regions of the globe. About 100,000 years ago, the first modern humans, sub-Saharan Africans, started moving inside Africa, and left this continent to migrate in the general direction of the Caspian Sea or due east toward India. Then, they moved toward the southeast, eastward, and westward to populate New Guinea and Australia, as well as Asia and Europe. Finally, they moved into the Americas. (From L. L. Cavalli and M. W. Feldman. 2003. The application of molecular genetic approaches to the study of human evolution. *Nature Genetics Supplement* 33:266–75, fig. 3. Reproduced with permission.)

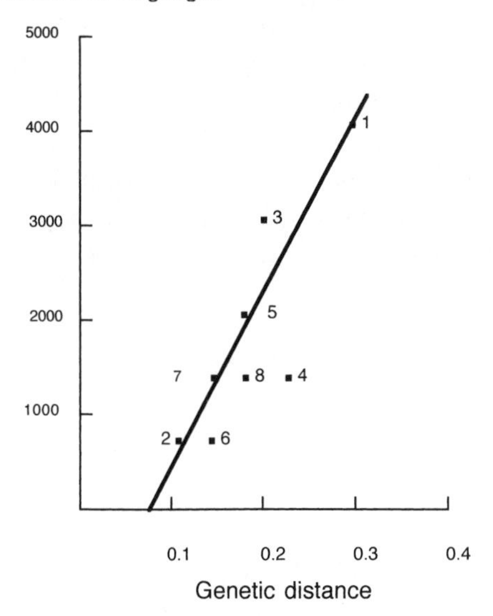

Figure 1.5 Correlation between the number of languages spoken by groups of human populations and the genetic distance that separates these groups. 1. All humans, 2. Africans, 3. Non-Africans, 4. Europeans / North Asians, 5. Southeast Asia / Australasia, 6. North Asia / Amerindians, 7. Mainland / Island Southeast Asia, and 8. New Guinea / Australia. The correlation coefficient is 0.91, which indicates very strong linearity of the relationship. This graph shows that, for example, when the whole world is considered (over 4,000 languages are tabulated in this graph), the genetic distance between the speakers of these 4,000 languages is much greater than the genetic distance separating speakers of North Asian / Amerindian languages (about 600 languages). (Redrawn from R. A. Foley. 1991. The silence of the past. *Nature* 353:114.)

Cavalli's suggested history and routes of expansion of modern humans in the last 100,000 years. In the area of linguistics, Cavalli demonstrated a strong correlation between language families and human genetic diversification. Figure 1.5 shows that increased genetic distance between human populations results in increased language diversification. This relationship is linear and is consistent with the idea that, as humans migrated and differentiated, their languages diversified concomitantly. Chapter 6 then describes Cavalli's most recent scientific efforts, including his work on DNA polymorphisms in the human male Y chromosome.

The next chapter is devoted to the Human Genome Diversity Project (HGDP) and some of the controversies surrounding it. The HGDP is a global study of human genetic diversity; it seeks not only to document this diversity but also to use this genetic information to trace human origins, map out prehistoric human migrations across the globe, and more recently, to make use of genetic information in the field of medicine. For this ambitious project, Cavalli aims to collect DNA samples from as many human populations as possible. A final chapter discusses the legacy of Cavalli's scientific contributions.

Chapter 2

From Medicine to Bacterial Genetics (1943–1960)

Luigi Luca Cavalli-Sforza was born Luigi Cavalli in Genoa, Italy, on January 25, 1922, the year Benito Mussolini and his Blackshirts marched on Rome to establish a fascist dictatorship. Cavalli's home country remained under Fascist and Nazi control until early 1945. By then, Cavalli had obtained his medical degree (in 1944) from the University of Pavia after having received his elementary and secondary education and spent his first year as a university student in Turin. One can see that Cavalli's early landscape was solidly Northern Italian, a region characterized by rich history and turmoil. As is the case for so many areas of Europe, Cavalli's birthplace suffered countless wars, invasions, and innumerable contacts and admixture with outsiders.

Genoa today is a major Mediterranean port and a city of 800,000 inhabitants. It is located in Liguria, which is part of the Italian Riviera. In prehistoric times, Liguria was inhabited by **Neanderthal** and subsequently **Cro-Magnon** people. It was at least partly Greek in the fourth century B.C., and was already in these days an important port. Ligurians, an Italo-Celtic people, warred with the Carthaginians and then with the Romans, who occupied the city until the fall of the Roman Empire. After this event, the Ligurians came under the influence of the Byzantine Greeks and, later, the Germanic Longobards and Franks.

The early and central Middle Ages proved to be a golden era for Genoa. Its port attracted international commerce, and its sphere of influence in the form of various dominions extended as far as the Black Sea. Genoa's archrival, Venice, located on the opposite side of the Italian boot, was also extending its own empire in the Mediterranean at the same time. This inevitably led to confrontation, war, and the ultimate defeat of Genoa in 1380. The city-state did, however, remain independent and flourished again as a shipbuilding port and a banking center. In 1797, Napoleon founded the Ligurian Republic, which became an integral part of France in 1805. After Napoleon's defeat, Liguria became in 1815 part of the Kingdom of Sardinia-Piedmont and was finally incorporated into the Kingdom of Italy in 1861.

Following that, all of Northern Italy was embroiled in social conflicts and the birth of socialism. Italy got involved in World War I, became fascist, made an alliance with Hitler's Germany, and was eventually defeated by Allied Forces in 1943 but was not entirely liberated until two years later. Being major enemy cities, Genoa, Turin, and Milan were extensively bombed by the Royal Air Force and the U.S. Army Air Corps during World War II. It is against this backdrop of dictatorship and war that Cavalli grew up and was educated.

In the 1920s and 1930s, pragmatic and technologically adventurous America was inventing electrical appliances at a very fast pace. Traditional Europe, having suffered the devastations of World War I, was less so inclined. Yet even in those days, some people foresaw the future of machines that would make the travails of European women a little easier. Cavalli's father was one of them. He became the European representative of an American firm, based in Syracuse, N.Y., that manufactured, among other things, electrical washing machines. He was successful and wrote a book entitled *La Spada dell'America* (The sword of America) that examined in detail sales tactics and advertising techniques in the United States. Following this, Cavalli's father was offered a professorship in advertising at Bocconi University in Milan, which he turned down (fig. 2.1).

In secondary school Cavalli studied Latin for eight years and classical Greek for five years. Learning these two dead languages was considered to be part

Figure 2.1 Cavalli with his father in 1936. (From the collection of L. L. Cavalli-Sforza)

of a well-rounded education. This practice continued for decades in Europe. Along with Latin and Greek, Cavalli learned French, strongly encouraged by his mother and closely monitored by his good teachers. He still prides himself today for his advanced knowledge of French. Later, between the ages of ten and nineteen, Cavalli also studied English and German, learning to speak and write them fluently. Cavalli mentioned that learning these languages was extremely important to him, saying that he could not overemphasize the difference that mastering these languages made to his life and career. As a result, he was able to travel with relative ease in many countries and was more readily accepted as a scholar internationally.

Contrary to twenty-first century Italy, in the 1920s Italy was still a deeply Catholic country. Mussolini had finally established a peace treaty between the state and the Church (thereby creating the Vatican), and Catholic schools, in particular those run by the Jesuits, were seen as centers of excellent learning. Even though Cavalli's family was not particularly religious, he was sent at age nine to a Jesuit school. There, he was forced to go to mass every single morning of the school year, an experience he found overwhelming. He has practically never gone to mass ever since. After this stint in Catholic school, Cavalli returned to public school where religious education was still part of the curriculum. In the religion class, mandatory in fascist times, when Cavalli was thirteen and in high school, the teacher gave his pupils a course largely dedicated to evolution. In doing so, he tried to convince his class that evolution was a wrong theory. Astonishingly, the teacher managed to convince Cavalli that the theory was right! Cavalli remembers asking the teacher at the end of the course whether believing in evolution was incompatible with remaining a Catholic. The answer was no (this is still the current view of the Vatican), in contrast to what is professed by fundamentalist Christian denominations.

Cavalli did not particularly enjoy high school, where the teaching was dull. He describes himself at this time (fourteen to sixteen years of age) as "very restless," and said his behavior was so bad that he was in danger of not being allowed to go on to the next year's class. So he stayed at home for one year, learning by himself the curriculum of the last two years of *liceo classico*, the equivalent of the last two years of high school. He then took the difficult high school equivalency exam and graduated at sixteen, three years ahead of time. Meanwhile, he had been forced by law to enroll in the Fascist Youth, an organization that meant to train the Italian young generations for good and enthusiastic participation in Mussolini's vision of the state. Such youth movements were not restricted to Italy at that time; Germany had its Hitler Youth, and other European countries where right-wing political parties were prevalent had similar organizations. As a young boy in Fascist Youth, Cavalli could not help but internalize some fascist doctrines. However, this same organization indirectly led him away from allegiance to these ideas since, through association with this group, Cavalli traveled to many other countries on study tours and so learned other ideas. To improve his English, his mother sent him to Exeter College in England (he was sixteen at the time). He told us that while there, he spoke about **fascism** to his English professors who, he said, "very kindly took me aside and explained about democracy."

Cavalli's parents wanted him to become an engineer, but he felt that the mathematical background he had received in secondary school was so insufficient that engineering was not an option. By and large, he could not decide what subjects to pursue at the university. He was interested in psychology, but he had also been intrigued by microscopes, which he had seen but not used when in high school. When registration day came, a floundering Cavalli thought he would enroll in medical school (in Italy in those days, one registered for six years of medical school right after high school), and prepared his registration documents accordingly. While standing in line, he discovered that the line next to his was for registration in the natural sciences, about which he had never heard. He thought that perhaps he should switch, but then decided against switching to avoid filling out a new application form. He reasoned that he could always switch later and so remained in the medical school line. This was the autumn of 1938, and in less than two years Italy would be at war. For Cavalli, his registration choice turned out to be an incredible stroke of luck because, contrary to all other university students, medical students were not drafted!

As mentioned, Cavalli started his medical studies at the University of Turin in the province of Piedmont. It so happens that Turin was, and still is, an important industrial center. This made the city a prime target for Allied bombers once Italy entered World War II, in 1940. After obtaining a good scholarship, Cavalli shifted to the University of Pavia and its Collegio Ghislieri. Pavia was a much smaller, much safer (and much prettier) medieval city not far from Milan. Medical school did not impress him very much. Attendance to lecture was optional and "you could learn more by just reading books, not necessarily the ones used in the courses." Hospital practice was not really mandatory and discipline was lax. In addition, the Germans were disrupting everything, even occupying Cavalli's college and forcing the students out of it.

However, Cavalli did more than study theoretical medicine at the university. In his third year of medical school, he realized that his quasi total ignorance of calculus, and a mental block against it, would not help him in a scientific career. He was fortunate enough to have the basics explained to him by an assistant professor working at the Polytechnic School of Milan. A little later, he even took a few formal courses in mathematics at the University of Pavia, but could not graduate with a math degree, given that by then he was employed full time by a pharmaceutical company. Nevertheless, Cavalli considers himself an "amateur mathematician," a modest qualification that many today would say underestimates his abilities. Mathematics would stay with him for all of his long career.

What's in a Name?

But before presenting this career, it is perhaps time to lift the veil of mystery shrouding Cavalli's double first name and hyphenated surname. Contrary to the Americans, the Europeans do not use a middle initial and practically never go by their middle name or both first name and middle name. Therefore, calling oneself Luigi Luca or even just Luca would be highly unusual for an Italian christened Luigi. As we know, some people in the United States dislike their first name, which they simply represent by its initial and then spell out their middle name. Europeans do not normally do that. Hence, calling oneself L. Luca is equally unusual. And then, there is this hyphenated name!

Cavalli's full name, Luigi Luca Cavalli-Sforza, impresses and mystifies most people. The hyphenated name suggests nobility, while the two Ls add a touch of romanticism. As mentioned, Cavalli was born Luigi Cavalli. In spite of the fact that his parents called him Gigino (an affectionate nickname meaning "little Gigi" or "Luigi"), Cavalli did not particularly like Luigi. But then, where does Luca come from? It turns out that Cavalli's genetics mentor, Adriano Buzzati-Traverso (mentioned in greater detail later), also disliked the name Luigi and suggested that Cavalli rename himself Luca. Thus, at age twenty, Cavalli became L. Luca Cavalli. In addition, Luca is a good acronym of *Lu*igi *Ca*valli. Many years later, at a conference on evolution held in Paris, Cavalli heard a speaker refer to him as "Luca," but not in a traditional sense. Indeed, this speaker revealed to him that Luca, when written LUCA, was the acronym for the unflattering expression "Last Universal Common Ancestor," a very simple organism indeed!

Next, what about the Sforza addition? It turns out that after his father's death, Cavalli was officially adopted at age twenty-seven by his childless maternal stepgrandfather, Count Francesco Sforza, and so became Cavalli-Sforza. Count Sforza must have been a formidable man. He descended directly from Duke Francesco Sforza (1401–1466), who ruled the Duchy of Milan during the early Renaissance and, not only that, he was also at one time director of the Bank of Italy in Milan. In that capacity, during the war, he hid Italian gold, thereby preventing the Germans from stealing it. Later, after World War II, he was governor of the Bank of Italy for all of Northern Italy. Therefore, Cavalli is of noble descent, although not biologically. Today, however, the Italian Republic does not recognize noble status or titles. Cavalli himself said to us about his noble ancestry: "I do not care about nobility."

This modification of names can be somewhat confusing, but it is not all that rare in Italy. For example, when we lived there once during a university sabbatical, our apartment building was located on a street without a name. Since that street was perpendicular to via Fuorimura, it became known by that name. However, some local inhabitants rejected that name and called the place Parco Tasso (Tasso Park) even though there was no park in sight! Similarly, a main street in Naples is known both as via Roma and via Toledo. These are just a few examples. On the other hand, people may not change their names that frequently, but multiple addresses and multiple telephone numbers are very common. Now that we have clarified this mystery, let us examine Cavalli's scientific career.

Work with Bacteria as a Medical Student and First Interest in Genetics

In 1940, as third-year medical students, Cavalli and his friend Giovanni Magni, who later also became a university professor of genetics, started their first research work. This was on the very nasty *Bacillus anthracis* (which causes anthrax) and later on *Pneumococcus* (which causes pneumonia). According to Cavalli, they wanted to provide accurate measurements of the virulence of these bacteria, something that, they were told, was impossible to do. These two had come up with the idea that, rather than simply measuring the percent of inoculated animals (mice) that *survived* the injection of a certain dose of pathogenic bacteria, also measuring the *time* of death of the animals would quantify and provide more insight into the notion of virulence. What is more, this dual measurement (dosage and time of death) might give hints as to what the genetic backgrounds of the experimental animals and bacteria contributed to the animals' sensitivity or resistance to the pathogens.

In these studies, one notices three factors: first, Cavalli (and presumably Magni) was thinking that the genetic composition of their test animals and bacteria were important in the development of an infectious disease. Second, they also began studies of mutations, induced by ultraviolet light and mutagenic agents, that affect bacterial metabolism and virulence. This was genetic thinking, as primitive as it may have been. Third, one finds the application of mathematical and statistical techniques to a biological problem, something that Cavalli understood was important, but that he was still very much in the process of mastering, as noted above.

Guidance in the field of bacterial virulence was completely unavailable in Italy at the time these specific studies were conducted, so Cavalli and Magni were told by their professors to go and visit researchers in Berlin and Frankfurt. The year was 1942; the season was summer. At that time, no part of Italy had yet been invaded by the Allies, and German and Italian troops were still holding practically all of North Africa (excluding most of Egypt), but Allied bombing of German cities had already begun. In 1942, Cavalli had gone to see the first bombed city, near Frankfurt. The battle of Stalingrad was a few months away. In short, in spite of some distracting carpet bombings, Germany and Italy could still hold on to the belief that they would win this war. And indeed, Cavalli reported to us that life in Germany was quasi-normal, with generous rationing—except for cigarettes, the allotted number being only three per day. These must have been trying circumstances for Cavalli and Magni, heavy smokers that they were.

Cavalli also reported seeing Jews identified, as we know, by a yellow Star of David, a sad sight, he said, adding that the Jews themselves looked miserable. On the subject of anti-Semitism in Italy during the Fascist era—in addition to its being odious, Cavalli told us that it was also very stupid. As had happened in Germany, all Italian university professors of Jewish descent had been fired by the Fascists. Since many of these ranked among the very best, this action left Italy deprived of a significant fraction of intellectual cadres. The whole Fascist-Nazi enterprise was an unbelievable waste of human talent, but this is another story . . .[1] To understand why Cavalli and Magni traveled to Germany in these troubled times, one must realize that, as young Italian scientists during World War II, they had no access to countries other than those belonging to the Axis powers, or to countries then occupied by German troops. In addition, Germany,

1. In spite of the openly anti-Semitic nature of the Fascist regime, by and large Italy did not impose a German-style "final solution" on its Jewish population. Rather than mass executions, the Italian authorities preferred job deprivation and exile to control the Jews. Accounts of these practices can be found in Carlo Levi's excellent novel *Christ Stopped at Eboli*. Today, Germans, including tourists, are not necessarily welcome in parts of Italy. We lived in the Naples area in 1989–90 and experienced this animosity firsthand. One of us (PFL) was repeatedly harassed because, seemingly, he has stereotypical phenotypical Germanic features. One time, in the town of Torre Annunziata, a few kilometers south of Naples, a motorcyclist swerved toward him, spat at his feet, yelled "Tedesco" ("German," in Italian), and sped away. This was a particularly stupid action, given that PFL was born in Belgium, a country that suffered much from German occupation in two world wars! Northern Italians, many of whom also have "Germanic" features, are not immune to discrimination. One of PFL's colleagues at the University of Naples, who is from Emilia-Romagna, and happens to be tall and blond, was once subjected to nasty verbal treatment (as a presumed "Tedesco") on a beach in Southern Italy. He answered back in the crudest possible Italian and immensely enjoyed the very puzzled look on his adversary's face. These anecdotes are not meant to reflect badly on Southern Italy as a whole. In fact, we like the place and its people so much that we have been back repeatedly in the past twelve years to visit our very good friends there.

even in those days, was a rich scientific powerhouse compared to Italy and other European countries. For example, while Cavalli and Magni were allowed only a few laboratory mice for their experiments in Italy, they had access to hundreds of experimental animals in Germany. Cavalli reported to us that their German host told them early in their stay: "Next Tuesday you can start an experiment with one hundred male mice, all weighing exactly sixteen grams. And you can repeat the same experiment two weeks later." To Cavalli and Magni, this was a baffling offer because, earlier in Italy, they could only procure about fifteen white mice at a time, and for this they had to visit many farms where the mice were raised. Not only that, they had to buy them with their own money!

After World War II, Cavalli stopped all contact with Germany for many years and only resumed any association with the country very recently. Still, he had formed a lasting impression of how well organized German society and German science were. Seeing this high degree of German organization compared to what he saw as the messy disorder in his homeland made him angry and mistrustful of Italy. What counts there, he said, is *arrangiarsi* (perhaps best translated as "finagling"). It was hard (and to some extent still is) to get things done in Italy without continual social maneuvering, conniving, and pulling strings, and the result, at least in science, is marked inefficiency. Cavalli has described Italy as a "scientific desert with a few oases."

As a result of their visit to Germany, Cavalli and Magni had received important scientific advice, including from the famous Soviet émigré geneticist Nikolai Timofeyev-Ressovsky, who Cavalli described as an "ebullient" man, and who was influential in later convincing Cavalli to become a geneticist himself. Curiously, Timofeyev-Ressovsky, after his arrest in Berlin by Soviet troops, shared a cell with writer Alexandr Solzhenitsyn in the Butyrki prison in Moscow in 1946. Meanwhile, Cavalli and Magni conducted several meaningful experiments in Germany and published three articles in German. This was not all that surprising for budding scientists in those days; back then, German was *the* language of science in continental Europe, not English as it is today. Years later, this work was published in English.

Then, in 1942, Cavalli's scientific life would be changed forever and his shift to genetics would become permanent. The German Afrika Korps and the Italian army had been defeated at El Alamein by the Eighth British Army under the command of Marshal Bernard Law Montgomery. Tens of thousands of German and Italian soldiers were taken prisoner. Not among them, thanks to asthma developed in the deserts of North Africa, was an artillery officer named Adriano Buzzati-Traverso, who had gone back to Italy in 1942, therefore avoiding capture. It turns out that Buzzati had learned genetics as a student at Iowa State University in the prewar years. Not only that, he had brought back with him many American textbooks on genetics and statistics.

These books, as well as Buzzati himself, proved to be a gold mine for Cavalli, who first met Buzzati while attending his lectures at Pavia (see fig. 2.2). Still a medical student, Cavalli learned genetics from his newly returned professor and started doing research on the population genetics of *Drosophila*, the fruit fly, and planktonic organisms.

Interestingly, Cavalli told us that Buzzati-Traverso considered science to be a sacerdotal endeavor of sorts: he believed a scientist should devote himself (there were practically no women scientists in Italy or elsewhere in those days) totally to his craft, and certainly never commit the indignity of getting married. Buzzati himself never did marry. Ironically, and unwittingly, however, Buzzati invited Cavalli to spend some time with him at his country home, where Buzzati's young attractive niece, Alba, was also visiting. Clearly turning a deaf ear to his mentor, Cavalli ended up marrying Buzzati's niece! Marriage, Cavalli told us, had always been high on his list of priorities. A year after meeting Alba, in 1944, Cavalli graduated from medical school. He and Alba were married in January 1946 (see fig. 2.3).

One would expect a person with a medical degree to start practicing medicine as soon as possible. In all countries in the world (except in the Soviet Union, for a few decades), medicine was and is seen as a highly desirable occupation, and a lucrative one to boot. Not so for Cavalli. In 1944–45, as World War II was still raging, Cavalli did practice some medicine, mainly in hospitals. This

Figure 2.2 Adriano Buzzati-Traverso (1913–1983), Cavalli's genetics teacher. (From the collection of L. L. Cavalli-Sforza)

Figure 2.3 Cavalli and his future wife, Alba, as fiancés in 1944. (From the collection of L. L. Cavalli-Sforza)

was not to his liking. As he told us, in those days physicians could do very little to ameliorate the fate of a patient suffering from a serious illness. Diagnostic techniques did of course exist but, often, appropriate treatments did not. In particular, the first antibiotic—penicillin—was unavailable in continental Europe at that time. Therefore, the role of the physician was restricted in many cases to deciding whether the patient would die soon or would die later, as Cavalli put it to us. Cavalli became a medical intern responsible for a 50-bed hospital ward. Given the general lack of proper therapies for most illnesses at the time, he told us, "I found medicine *so* depressing. I didn't like it." He did acknowledge, however, that, later, the field of medicine became so much better, and saved his life multiple times.

Having been exposed to basic research during his medical studies and enjoying it, Cavalli quickly abandoned medical practice. One skill he learned that would stay with him (and that also proved useful in his future research in human genetics) was the ability to perform painless venous extractions of blood to study proteins in the harvested fluid, a key element in the understanding of genetic diversity among human populations, as we will see later in this book. However, human genetics was still several years away on Cavalli's horizon. Before embarking on the study of the origin and migrations of humans, Cavalli would spend several years revolutionizing bacterial genetics.

Postwar Activities

As a recently graduated medical doctor, and having discovered that there were no university jobs available, Cavalli got a job in 1945 at the Istituto Sieroterapico Milanese (Institute of Serumtherapy of Milan), a nonprofit pharmaceutical company (fig. 2.4). There he had to do lots of statistics, which enhanced his mathematical skills. At this institute he also was able to convince his director to let him conduct basic experiments on the sensitivity of bacteria (**Escherichia coli**, a common inhabitant of the human intestine) to a variety of **mutagens** known to induce mutations in other organisms—agents such as ultraviolet radiation, X-rays, and certain chemicals. This work was done in collaboration with his colleague Niccolò Visconti and was published in Italian in 1948. Their work was highly quantitative in the sense that they did not simply "blast" their bacterial cultures with toxic agents; they took great care to measure these agents' effects in the different phases of the growth cycle of their bacteria. They also ran statistical calculations on their results, and they worked on the physiology of an *E.coli* mutant srain, called *B/r*, that was at least ten times more resistant than regular *E. coli* to ultraviolet irradiation and X-rays.

When reading these articles, one cannot avoid noticing the quality of the "Discussion" section of their work. The authors took great care to incorporate

Figure 2.4 Looking at a bacterial plate (Italy, c. 1948). (From the collection of L. L. Cavalli-Sforza)

alternative explanations to their own conclusions, showing an excellent understanding of the transitory nature of scientific explanations. This attitude has prevailed throughout Cavalli's career and can be contrasted nowadays with all-too-frequent hurried articles that find their way to press in our modern, fast-paced, and cut-throat scientific world.

Before proceeding, it is useful to raise a question: Why is it important, in genetics, to isolate and study **mutants**? After all, the usual connotation of the word is "crippled or unfit." And these definitions are largely correct in some contexts. The very simple answer to this question is given by raising another: How does one know what a gene does in an organism if this gene is always functioning perfectly well? But then, organisms contain thousands of genes that all act in concert to keep biological functions working at top performance. How is one to dissect all these genetic pathways and understand what individual genes do? Altering by mutation one gene at a time will often tell us what this gene does when the destruction is followed by a specific failure in the organism's performance. For example, mutating a bacterial gene responsible for the making of vitamin B1 will make the bacteria dependent for life on an external supply of this vitamin, something that can easily be observed and studied in the lab. Supplementing the broken gene's function with the appropriate nutrients will then reveal what the intact gene was supposed to do in the first place. Or else, as in Cavalli's results with genetic resistance to ultraviolet light and X-rays, one can research the causes of this resistance (we now know that enhanced DNA repair mechanisms are at work here). Thus, the mutated genes and their altered functions can be investigated because they are now present in mutant organisms whose phenotypic properties are easily identifiable. So one important aspect of genetics is to mutate a gene in order to understand what it does when it is not mutated.

To understand the next step in Cavalli's career, it is necessary to digress somewhat and shift our attention to what was going on in the United States in the realm of bacterial genetics while Cavalli was working on his *E. coli* mutants. As we shall see, other mutants were instrumental in the studies that spurred Cavalli's interest and steered his career for a good ten years.

Sex in Bacteria Is Discovered in the United States

The study of genetics invariably involves sexual crosses—that is, matings between individuals of opposite sex. The expression of the parental genes are then examined in the offspring of the crosses. Starting in the mid-1940s, many geneticists began to show an interest in bacteria. Up until that time, plants and *Drosophila melanogaster*, the fruit fly, and to a certain extent mice, had been used as successful genetic tools. On the other hand, plants, mice, and flies are complicated multicellular organisms, surely much more complicated than simple unicellular bacteria. What is more, the fact that bacteria have genes made of DNA had been demonstrated in 1944 by Oswald T. Avery and colleagues. Indeed, this was the first convincing demonstration that DNA had any genetic properties at all, a view not held at the time by most scientists, who preferred to think that proteins constituted the genetic material.

Why not, then, use bacteria as genetic tools? They multiply very rapidly and abundantly, can be grown in small volumes of nutrient broth, and need only small incubators for growth, rather than vast fields or countless fly bottles. There was one big problem: bacteria seemed to be unwilling to engage in sexual reproduction, and even the idea of male and female bacteria seemed ludicrous to many.

Enter Joshua Lederberg and Edward Tatum, working at Yale University. In 1946 these two published the first evidence that bacteria can indeed reproduce sexually. For this, they also used the bacterium *Escherichia coli* (*E. coli*), then recently made a popular tool for genetic experiments. Here is how they did their first investigation. Lederberg reasoned that one way to show sexual mating in bacteria was to use partners that had lost the ability to grow on minimal medium. Minimal medium is broth that contains salts and a sugar dissolved in water. No organic compounds other than sugar—used as a source of carbon and energy by the cells—are added. Normal *E. coli* cells grow perfectly well in this type of medium because they can use the sugar and the various salts to manufacture all the organic compounds (such as amino acids, DNA bases, and vitamins) they need in order to divide. However, when the genes responsible for the manufacturing of these organic compounds are mutated, mutant cells can no longer grow in minimal medium because these organic compounds are no longer made.

Tatum had generated a whole collection of mutant *E. coli* strains deficient in a variety of amino acids and vitamins. Lederberg, meanwhile, had reasoned that if *E. coli* cells can mate, and thus exchange genes, the normal genes from one parent could supplement the defective genes of the other parent. In one experiment among many, he and Tatum used as one parent strain Y 10, a strain unable to manufacture the amino acids threonine and leucine, and the vitamin thiamin. Due to this triple mutation—indicated as thr$^-$ leu$^-$ thi$^-$—Y 10 cannot grow in minimal medium. The other parent was strain Y 24, another triple mutant, unable to make the amino acids phenylalanine and cysteine, and the vitamin biotin. This strain is designated as bio$^-$ phe$^-$ cys$^-$ and is also unable to grow in minimal medium. The experiment consisted in incubating together strains Y 10 and Y 24 in minimal medium, containing none of all the necessary amino acids and vitamins. If genes were transferred from Y 10 to Y 24 or vice versa, the normal genes of one or the other should supplement the mutated genes, and the cells resulting from that cross should be able to grow in minimal medium. This is exactly what happened (fig. 2.5A). For each one million mutant cells grown together, about one cell was able to grow in minimal medium and thus contained normal thr, leu, thi, bio, phe, and cys genes. This meant that the normal thi$^+$, leu$^+$, and thr$^+$ genes of Y 24 had been transferred to the bio$^+$, cys$^+$, phe$^+$ Y 10 parent, thereby "repairing" the three thi$^-$, leu$^-$, and thr$^-$ mutations in Y 10 (fig. 2.5B).

Further studies by Lederberg and Tatum confirmed this phenomenon with other sets of genes. Bacterial sex—that is, DNA exchange between bacterial cells—had been demonstrated. We know today that the genes from the donor parent enter the other parental cell and recombine with the host genes to form a new recombinant chromosome. This happens because the two chromosomes, now present in the same cell, break and rejoin between the two sets of triple mutations (fig. 2.5B and C). This phenomenon is known as **crossing over** or **recombination**.

It was later demonstrated that *immediate physical contact between the two parents was necessary*, further reinforcing the idea that the genetic results were due to mating followed by gene transfer. In addition, this newly discovered

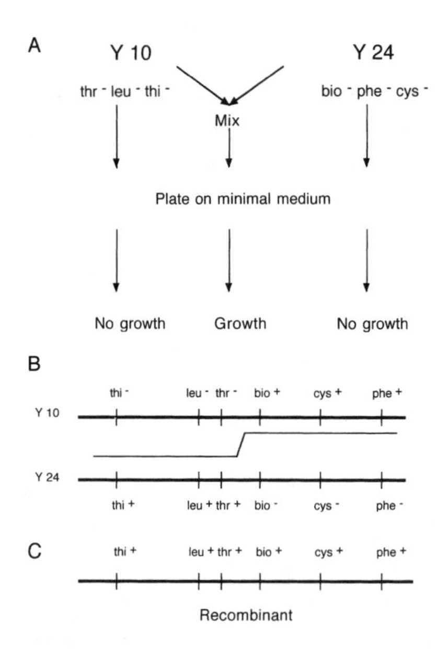

Figure 2.5 The original Lederberg and Tatum experiment that demonstrated sex in bacteria. A: The genotypes of the two parent strains (only mutant genes are shown). B: Crossing over between the two chromosomes temporarily present in the recipient cells. C: Recombinant chromosome obtained after crossing over shown in B. The recombinant chromosome no longer contains any mutations, which is why these cells can grow on minimal medium.

phenomenon allowed scientists for the first time to determine the precise order of the genes present in the bacterial chromosome. Such gene mapping, as the concept is called, had been performed decades earlier in the fruit fly *Drosophila melanogaster* and some crop plants and mice, but this was possible because, of course, flies, mice, and plants reproduce sexually. Lederberg and Tatum had thus shown that gene mapping via sexual crosses could be extended to bacteria. Their observations had opened up a brand new field—the study of bacterial genetics via **conjugation**, or mating. Cavalli would not wait long to join in.

Meeting an English Genius

Italy in the late 1940s was a lonely place for a researcher interested in the new science of bacterial genetics. Scientists there (and elsewhere) interested in the study of bacteria called themselves bacteriologists and generally showed great disdain for the study of bacterial genes. Many still thought—erroneously and in spite of some good evidence to the contrary—that bacteria probably did not even have genes; bacteria were rather seen as bags of proteins that somehow managed to divide. Thus, Italy was no fertile terrain for a young and ambitious bacterial geneticist.

In 1948, Cavalli obtained a scholarship to spend some time in an English lab, and from there he went to Stockholm to attend the International Congress of Genetics and present his results on the effects of nitrogen mustard and radiation on bacteria (see below). There, he introduced himself to a great man, Ronald A. Fisher (later, Sir Ronald), professor of mathematical and statistical genetics at Cambridge University. Fisher was one of three geneticists (the other two being the American Sewall Wright and Britisher J. B. S. Haldane) considered the architects of the neo-Darwinian synthesis, the theory that merged genetics and evolution by natural selection. Fisher was also a formidable statistician and indeed the father of modern statistics and experimental design, a fact that much impressed the now quantitatively minded young Cavalli. In addition, Ronald Fisher was a skilled experimentalist, performing his own crosses with mice, making the observations, and recording the results personally. Curiously, Fisher had very few mathematically inclined students, perhaps because genetics did not appear sophisticated enough to "real" mathematicians. Be that as it may, in Cavalli's words, "Fisher was a true genius."

To Cavalli's surprise, Fisher immediately offered him a job at Cambridge University. Why? According to Cavalli, Fisher must have read and understood better than anybody else the 1946 Tatum and Lederberg article describing recombination in *E. coli.* Chromosome recombination, crossing over, and gene mapping were very high on Fisher's list of interests, but in

Figure 2.6 Cavalli (*right*) with Sir Ronald Fisher (*middle*) and unidentified person (Italy, 1954). (From the collection of L. L. Cavalli-Sforza)

those days his Cambridge lab was mostly doing research with mice. Mice are of course not the easiest biological material to work with, and Fisher—statistician that he was—was impressed by the enormously large numbers of bacteria one could grow in very small volumes. Very large numbers are statistically very useful, bacteria grow quickly and in large numbers, and Cavalli knew bacteria. This simple syllogism explains why Fisher wanted Cavalli in his lab (see fig. 2.6).

Cavalli was extremely excited to be offered a job in this Mecca of science and immediately accepted. He moved with his family—now consisting of his wife Alba and their young son Matteo—to Cambridge in October 1948. Overall, Cavalli's view of England was very positive. He praised the British people for their eccentricity, humor, and "dignity in adversity," although he thought that the British might be "too reserved," especially at Cambridge. More to the point, science in England he saw as highly efficient.

Cavalli has interesting recollections of the thinking of the great Ronald Fisher, such as the following: once they were discussing languages and Cavalli mentioned that he thought French was more logical than English. "But English," Fisher replied, "is better for science because it has the right degree of ambiguity." Giving this some thought, Cavalli agreed; as he told us, "In science you often need to express the right level of uncertainty."

Sadly, even though World War II had been over for more than three years, Great Britain, still reeling and nearly bled to death by the war effort, devastated and having suffered immensely from straitened economic circumstances both during and after this conflict, was still experiencing rationing. Yet spirits were high. Cavalli was asked to teach a course on microbial genetics, which he thinks may have been the first in the world. He also set up a bacterial genetics laboratory that was carved out of an old abandoned tearoom about half the size of Cavalli's eventual office at Stanford.

There, Cavalli quickly repeated, confirmed, and extended the results of Lederberg and Tatum after receiving their *E. coli* strains. It is Cavalli who initiated what was to become an enormously fruitful collaboration with Joshua and Esther Lederberg. A letter dated September 9, 1948, mailed from Cambridge, England, to Joshua Lederberg shows Cavalli requesting Lederberg's *E. coli* strains and wishing to remain in contact with his American colleague. In line with historical events, this letter was signed "Luigi Cavalli," and not "Luca Cavalli-Sforza."

But simply repeating (and confirming) the Lederberg-Tatum experiments is not all Cavalli did. As mentioned, gene transfer between bacteria occurred at the appallingly low frequency of about one transfer per one million cells used in the mating process. This is like crossing one million fruit flies only to end up with a single successful mating that produces progeny. If that had been the case, fruit flies would never have been chosen as material for genetic studies. This state of affairs in regard to *E. coli* was, then, far from satisfactory if one wanted to map the many genes suspected to be harbored by these bacteria. One needed to improve the frequency at which *E. coli* cells mated or, to use modern terminology, conjugated. This feat was achieved by Cavalli in 1949.

In these "classical" days of genetics, it was known that a variety of agents (such as ultraviolet light and X-rays, as well as certain chemicals) could induce heritable changes in the biological properties of living organisms, including bacteria. In 1948, working at the Istituto Sieroterapico in Milan, Cavalli had published a mathematically oriented article on the effects of these agents on the viability of *E. coli* (then called *Bacterium coli*). As mentioned above, these agents are called mutagens because they induce changes (mutations) in the genetic material of living cells.

These changes are random because mutagens are potentially capable of altering any gene in an organism. In other words, if one wants to isolate a mutant with specific new properties after mutagenesis, one must rely on two factors—a good dose of luck and a lot of hard work. For this reason, a search for mutants is often referred to as a mutant "hunt." Cavalli proved to be a good mutant "hunter" (but not so much so at hunting animals, which he once tried in Austria—and then rejected with disgust after killing a roebuck); but he successfully isolated an *E. coli* mutant able to conjugate up to 10,000 times more frequently than the strains used by Lederberg and Tatum. To achieve this, he used the horrendously toxic chemical mutagen nitrogen mustard, a compound whose mutagenic properties had been discovered by Charlotte Auerbach just a few years earlier. This compound is also known as yperite because it has the dubious honor of being a gas agent used by the Germans against British and Canadian troops defending the beautiful medieval city of Ypres (aka Ieper), in Belgium, during World War I. Even though Ypres was not taken by the Germans, in spite of the gas attacks, the city was completely leveled by heavy artillery bombardments in 1915. To understand why such a horrible concoction was used later to do inoffensive experiments, we must realize that, in the 1940s, the choice of mutagens was extremely restricted and further, by their very nature, *all* mutagens are toxic. Nevertheless, modern geneticists prefer to avoid nitrogen mustard because it is potentially dangerous.

Thus, Cavalli had isolated in 1950 what he christened, in English (in the midst of an article otherwise written in Italian), an **Hfr** strain, which stands for *H*igh *fr*equency of *r*ecombination. And indeed, this is exactly what this mutant does: it performs at high frequency the phenomenon described in figure 2.5. A few years later, this strain was rechristened by others *Hfr* C, where C stands for Cavalli. *Hfr* C can be found in the mutant collections of many laboratories and is still in use today, more than half a century after its isolation. This was Cavalli's first great contribution to the science of genetics. But what was it that *Hfr* C had acquired through mutagenesis that other *E. coli* strains did not possess? Why did *Hfr* C mate so easily and transfer its genes to other cells when other strains did not? Solving this problem meant first getting a better grasp of the nature of sex in bacteria. Cavalli, together with Joshua and Esther Lederberg, eventually clarified the mechanism of bacterial sex in 1952–53, working on different continents but publishing together.

Meanwhile, Cavalli, still in England at Cambridge, had turned his attention to a special category of genes called *polygenes*. It is a hallmark of Cavalli's career that he often tackled different research projects simultaneously. This makes the task of the biographer more complicated, but as Cavalli told us, "research is more efficient that way." Also, Cavalli normally used the strategy of carrying out two different investigations at the same time: a more exciting one with low probability of success, but giving high satisfaction if successful, and a safer, less stimulating one, with greater hope of return.

What are polygenes? The prefix *poly-* means several. Hence, **polygenic** traits are traits that are under the control of several genes. Familiar examples in humans include skin color, early-onset diabetes, and autism, the latter studied by Cavalli and colleagues in the 1990s. Such traits are complicated, not only because they are caused by many genes (three or four variants in the case of skin color and more than ten in the case of autism), but also because the expression of these genes is influenced by the environment. For example, skin color depends on the amount of exposure of our skin to solar ultraviolet radiation. Similarly, the severity of a diabetic condition varies with diet. Many genetic diseases in humans—such as hemophilia, cystic fibrosis, and sickle-cell anemia—are caused by defects in single genes. However, most of these diseases are quite rare. Conversely, predisposition to mental illness, cancer, high blood pressure, heart problems, and several other diseases are under the control of many genes. Our understanding of the genetic basis of these diseases is presently very limited, but the deciphering of the human genome promises great advances in this area of medicine.

Cavalli started working on polygenes, using bacteria (*E. coli*) as an experimental system. Together with his colleague G. Maccacaro (a fellow from his University of Pavia days whom he had invited to join him at Cambridge), he demonstrated that bacterial resistance to the antibiotic chloramphenicol (then called chloromycetin) was best explained by changes having occurred in at least five genes. The research showed a remarkable result. When a highly resistant strain, obtained by prolonged cultivation in increasing concentrations of the antibiotic, was crossed to a sensitive strain, chromosomal recombination separated the high resistance into a series of steps. When a highly resistant strain was crossed to another highly resistant strain (obtained by an independent process of selection for high resistance), the progeny showed a considerable variation of resistance, from full sensitivity to moderate, to high resistance, but never higher than that of the parental strains. The interpretation was as follows: the first mutant selected was of course resistant. However, later mutations, added by continuing the process of adaptation, need not have been resistant by themselves; they simply modified the resistance of the first, increasing it. Experiments separating by recombination the first from the later mutations, or combining those acquired in different independent processes, showed many different interactions between different mutations. Some mutations increased resistance in an additive manner, others did not, or even decreased it.

And nobody paid the slightest attention to these results! Why? The reason is that, in the 1940s and for decades later, bacterial genetics involving traits under the control of single genes (such as streptomycin resistance, for example) was complicated enough to deter researchers from studying complex traits under the control of polygenes. Then, once the idea of studying simple traits took hold, everybody forgot about polygenes in microorganisms. Astonishingly, colleagues of Cavalli's at Stanford University are currently revisiting polygenes in microbes—more than fifty years after he published his first results! Cavalli is quite happy to note that these researchers have found excellent evidence that a trait in yeast—resistance to heat—is under the control of four genes, not just one. These genes may be involved in respiration (the ability of yeast to use oxygen) or in the formation of the cytoskeleton (the protein scaffold that gives cells their shape). Some of these genes are located very close together on chromosomes, and their interactions are quite complicated. Their analysis was possible only in an organism like yeast, which lends itself especially well to a highly refined genetic analysis. Hopefully, the study of complex traits in microorganisms will continue, because these organisms could be used as simple models for the behavior of complex traits in humans. So far, the results with yeast polygenes show that the study of complex diseases such as hypertension, cardiopathies, arthritis, asthma, and several major psychiatric disorders will be much more demanding than one would hope.

Back to our story. The year is 1950 and Cavalli is getting ready—reluctantly—to leave England and return to Italy. His second son, Francesco (now a movie director and writer), had just been born in Cambridge, and work with *Hfr* and polygenes was going well. What had motivated Cavalli's move back to Italy? Simply, Fisher had not been able, in spite of forceful argumentation, to convince funding agencies that bacterial genetics had any future, and this program was in danger of being eliminated. Unsure of his future in England, Cavalli accepted an offer to return to Italy. Also, Cavalli had felt isolated in Cambridge, with very few people around to discuss the topic of bacterial genetics. So he thought it better to return to the scientific desert from whence he had come, as there seemed to be at least some chance of future progress there.

Back to Italy: Solving the Riddle of Sex in Bacteria

Cavalli resumed his old position at the Istituto Sieroterapico in Milan. The work he did there in the next several years, in collaboration at a distance with the Lederbergs, and against a background of work by others, would revolutionize our understanding of bacterial sexuality and lead to discoveries that still have an impact today in the fields of genetic engineering and genome sequencing.

The concepts that we explain below are not simple; James Watson, just before he helped discover the DNA double helix, traveled to Milan to try to understand what Cavalli and the Lederbergs were doing. He found Lederberg's thinking "rabbinical" (as he called it in his 1968 book *The Double Helix*), an unjustified comment, but one attesting to the complexity of the problem. After all, if even a future Nobel laureate had trouble understanding these genetic experiments, they must indeed have been arcane.

But then, if the word "rabbinical" was meant as a joke, the joke was later on Watson himself. In *The Double Helix* he mentions that he went to Milan to get information "straight from the horse's mouth." Several years after the book was published, Watson chaired one session at a scientific conference held at Stanford University, where Cavalli (which means "horses" in Italian) was to be a speaker. Cavalli told us that, after being introduced by Watson, and referring to his book in his reply, he asked him if he was aware that his name indeed meant "horses"—to which Watson replied, "Oh, I didn't know that Sforza meant horses in Italian!" A linguist Jim Watson is not. Now back to science.

In what follows, we mention only Cavalli's name, but the reader should understand that the articles discussed were written in collaboration with the Lederbergs. Cavalli considers that his discovery of the *F* system in *E. coli*—which determines sex in this organism—is his greatest achievement in bacterial genetics (it is Joshua Lederberg who actually came up with the *F* designation). First, Cavalli demonstrated that, like most living species, *E. coli* bacteria come in two different sexes, but with a twist. By mating different *E. coli* strains, he—and the Lederbergs—independently realized that some crosses were fertile and some others were not. Remember that a fertile cross is one in which the genes of one parent enter the other parent and gives rise to the production of recombinant progeny, as in figure 2.5.

Let us now define an infertile (or sterile) cross as follows: *F–* × *F–* is sterile, in which *F* stands for fertility. The minus sign indicates that two **F–** strains cannot mate; they are sexually incompatible and produce no recombinant offspring. Fertile strains are called **F+**. Eventually, it was shown that *F+* × *F–* crosses were the most fertile, but a cross between two *F+* strains was also fertile.

At this point, one might think that this all makes some sense because, after all, in humans for example, it takes two members of the opposite sex to produce progeny. But the twist with *E. coli* is that an *F+* × *F+* cross is also fertile, as we saw above. This is the equivalent of a fertile mating between two men or two women! Another strange observation was that after an *F+* × *F–* mating, all the progeny cells had become of only one sex, *F+*. In humans, the equivalent would be that a man mating with a woman would transform her into a man, and all their children would be boys! Clearly, sex in bacteria is of a different nature: something odd (to us, if not to the bacteria) is going on in these experiments.

Cavalli's interpretation was that *F+* cells contained a factor of unknown composition, called F, that made these cells potential donors of genes to recipient *F–* cells, which do not contain this F factor. In a sense, this F factor behaved like a virus, because it was transmissible to *F–* cells by simple contact with *F+* cells. However, contrary to a virus, the F factor was never found outside bacterial cells. Thus, F was not a virus. But then, what about the famous *Hfr* strain isolated earlier by Cavalli? He realized that *Hfr* had originated from a mutagenized *F+*, which explained very well why an *Hfr* × *F–* cross was fertile. Thus, *Hfr* was a special form of *F+*. But then, whereas an *F+* × *F–* cross yielded 100 percent *F+* progeny, this number was 0 percent in an *Hfr* × *F–* cross. What had happened to the F factor that had suddenly become cryptic in an *Hfr*? It must not have been completely lost, since *Hfr* cells could donate their genes to *F–* cells at high frequency. A first hint came to Cavalli when he realized that an *Hfr* cell could spontaneously revert to a regular low frequency of recombination *F+* cell. The reverted *Hfr*s then behaved like normal *F+* in crosses with *F–* cells—they gave 100 percent *F+* in the progeny. Thus, an *F+* could become an *Hfr* (through mutagenesis), and this *Hfr* could spontaneously become an *F+* again. In other words, the F factor could exist in at least two forms in *E. coli* cells.

A year later, Cavalli, in collaboration with John Jinks of the University of Birmingham, solved the problem by mapping the position of the F factor on the *E. coli* chromosome. They had observed that the previous result that *Hfr* × *F–* gave 100 percent *F–* progeny was not entirely correct. In fact, by using refined techniques, they observed that an *Hfr* × *F–* mating gave 99.7 percent *F–* and 0.3 percent *Hfr*. In other words, it was possible to retrieve recombinant progeny containing the F factor in a "bound" state—that is, the F factor physically attached to the *E. coli* chromosome. Their results showed that the F factor in *Hfr* mapped between the gene responsible for the metabolism of the sugar maltose and resistance to the antibiotic streptomycin, and the gene responsible for the metabolism of the sugar galactose, with some uncertainty as to the exact point of integration. In addition, and based on their analysis of the frequency of transmission to *F–* of the genes located to the left and to the right of the *Hfr* locus, they concluded that, somehow, *Hfr* corresponded to a chromosomal breaking point defining a left chromosomal "arm" and a right chromosomal "arm." They also concluded that the chromosome of *E. coli* was linear, a rod of sorts.

It turns out that this last conclusion is the only wrong one in a long list of Cavalli's revolutionary correct discoveries. We now know that the chromosome is *transferred* to *F–* cells in a linear fashion. Elie Wollman and François Jacob, two French researchers who repeated and extended Cavalli's work, demonstrated later, in 1961, that the *E. coli* chromosome is actually circular. Further, they refined the location of the point of integration of F in *Hfr* C and showed that the breaking point was in fact situated between the genes for the metabolism of lactose and galactose, respectively. When we asked Cavalli whether he had perhaps suspected the circularity of the chromosome (he was oh so close . . .), he candidly answered "No."

By 1953, the work of Cavalli with the Lederbergs and Jinks had demonstrated the following principles:

1. Some *E. coli* cells contain a fertility factor F rendering them able to transfer at high frequency the *F+* behavior itself but also, and much more rarely, chromosomal genes to an F factor-less *F–* recipient.
2. When an *F+* is converted into an *Hfr* strain, the F factor is able to mobilize chromosomal genes to an *F–* recipient at very high frequency. This is possible because the F factor becomes *integrated* at a specific spot in the *E. coli* chromosome.

3. The point of integration of the F factor in an *Hfr* breaks before chromosomal genes are transferred into the *F–* cells. The integrated F factor is transferred to the *F–* only rarely.

We know today that all these interpretations are correct. The reason why *Hfr* cells transfer their chromosomal genes at high frequency is because, there, the F factor integrates within the bacterial chromosome at high frequency. The integrated F factor then "drives" the linearized *E. coli* chromosome (or the chromosome broken at one single point) into the recipient cells. In regular *F+* cells, this integration occurs rarely, which is why most of the time these cells transfer only the F factor to recipient cells and only very rarely transfer chromosomal genes.

What is the nature of this F factor? It took sixteen years to answer this question. Other researchers eventually showed that F is simply a piece of circular DNA, a minichromosome, that coexists with the chromosome in an *E. coli* cell (fig. 2.7). In its free state, F DNA (that is, the DNA of the F factor)

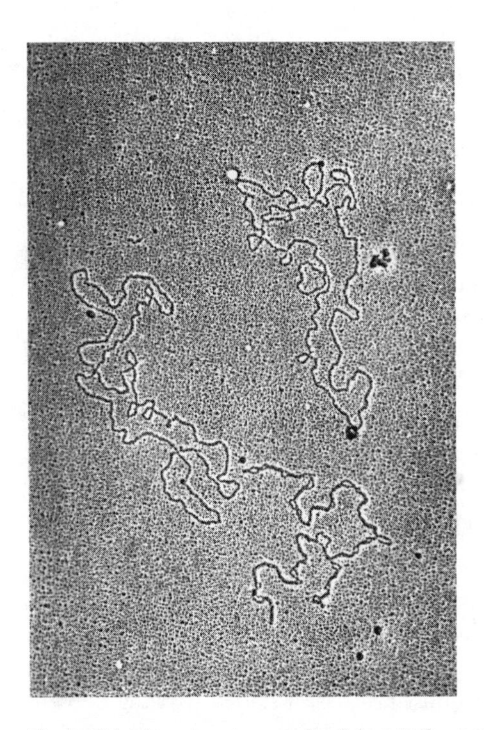

Figure 2.7 Electron micrograph of three F factor-like extrachromosomal circular DNA molecules from *E. coli*. One of the three molecules is twisted on itself (supercoiled). (From P. F. Lurquin. 2001. *The Green Phoenix: A History of Genetically Modified Plants*. New York: Columbia University Press. Reproduced with permission.)

transfers itself, via a conjugation tube, to an *F–* recipient, making it *F+*. In an *Hfr*, F DNA is integrated, physically part of the chromosome, and no longer free. It can mobilize the entire chromosome, and all the genes it carries, into an *F–* recipient. Most of the progeny of this cross are still *F–*, because the conjugation tube through which the chromosome and the integrated F travel is fragile and breaks before F has had a chance to penetrate the *F–*. Therefore, integrated F, as in an *Hfr*, penetrates the recipient *F–* last.

All modern general genetics textbooks describe the *Hfr / F+ / F–* experiments at some length. Most of them also mention Cavalli's name. The reason for this is that Cavalli, even though he did not *discover* bacterial sex, did explain in great detail just *how* bacteria interact sexually. The reader knows now that bacterial conjugation is a highly sophisticated process that took great intellectual effort to solve. Many a genetics professor loves to teach the portion of a general undergraduate genetics course where Cavalli's results are described. This is because the logic of these experiments is so elegant and implacable. Student reaction is more mixed; many prefer to contemplate the less abstract model of the DNA double helix because, they think, this is a material object they know intimately and which must therefore be more "real." One could disagree with them, but this a psychological question, not a scientific one.

In addition to being a genetic classic, the *F* system is used today to clone large pieces of DNA in structures called *bacterial artificial chromosomes*, or "BACs." These BACs have been used in the sequencing of the human and other genomes. They are also used to isolate genes from plants and mammals for the purpose of genetic engineering. As we know, deep controversies surround the genetic modification of plants, animals, and perhaps even humans. However, Cavalli and the Lederbergs cannot be accused of having caused the birth of genetic engineering. Nevertheless, the consequences of his "rabbinical" work, done more than half a century ago, are still with us. In a very indirect fashion, Cavalli's early science has also met society.

The Challenge from Ireland

It can be lonely at the top in science, but this loneliness is rarely of long duration. Even though the question of bacterial sex might have seemed like the ultimate in Byzantine thinking to the general public, another scientist, William Hayes, a medical doctor himself, had also started working on conjugation. Cavalli had met Hayes in England and had taught him how to do crosses with *E. coli*. He had also provided him with crossable strains. Hayes, an Irishman working at the University of Manchester, demonstrated, by using inhibitors and *F+* × *F–* crosses (he did not use the designation "F," which was first published by Cavalli and the Lederbergs), that one of the two strains was a gene donor, whereas the other one was a gene recipient. Cavalli had already demonstrated that by showing that *F–* × *F–* crosses were sterile. Hence, *F–* cells could not be gene donors. Through mail exchange, Cavalli and Hayes found out that they had been working on exactly the same problems and obtained very similar results! They then decided that the best thing to do was to publish their results separately but simultaneously. The two articles came out in 1953, in the same journal, and mark the discovery of the sexual differentiation in *E. coli* being caused by a transmissible agent, the F factor.

Incredibly, shortly after Cavalli had isolated his *Hfr* C, Hayes isolated another *Hfr* strain (later called *Hfr* H). Contrary to Cavalli, Hayes did not use mutagens to produce his *Hfr* from an *F+* strain; he simply rechecked the properties of an old *F+* strain that had been lying around in his refrigerator for a while. He discovered that his own old strain had spontaneously mutated to become an *Hfr*. In fact, the properties of *Hfr* H were indistinguishable from those of *Hfr* C in crosses with *F–*, except for two things. First, the breaking point of the transferred chromosome—in other words, the point of integration of the F factor—was different from the equivalent point in *Hfr* C. In *Hfr* H, this point was located between the genes responsible for the making of vitamin B1 and those responsible for making the amino acids threonine and leucine. Second, the direction of gene transfer from *Hfr* H relative to *Hfr* C was reversed. This observation, and similar ones made with later *Hfr*s, allowed Wollman and Jacob to conclude that the *E. coli* chromosome was circular.

But there was some disagreement between Hayes and Cavalli as to how exactly gene recombination in *F+* × *F–* and *Hfr* × *F–* crosses worked. For Cavalli and the Lederbergs, the *F+* donor transferred *all* of its genes into the recipient *F–* at once and, subsequently, many of these genes were eliminated, thrown out, whereas just a few remained in the *F–* to eventually recombine with its chromosome. For Hayes, the *F+* was only able to transfer *some* of its genes at any one time to the *F–*, following which *all* of these few genes recombined with the *F–* chromosome. In this case, Hayes proved to be right. (One might say: "Of all the rabbinical, Byzantine ratiocinations . . . ") But then again, these scientists were building the science of bacterial genetics, and no question seemed too futile to be tackled. Some of this contorted thinking was not lost on other geneticists, however. According to Cavalli, in a published article, one of them made the remark (in a jocular manner): "How can you believe this bacterial sex thing, knowing it was discovered by a Jew, an Italian, and an Irishman!"

Finally, scientific controversy is the grist of the mill of science. It forces scientists to go back to the bench and do experiments to try to prove their point or disprove their competitors' claims. And one never knows what treasures will be found in the midst of new data. As for Hayes, he was not always right either. In 1953, Cavalli was invited by him and Watson (by then of DNA double helix fame) to coauthor an article stating that *E. coli* had three chromosomes. Cavalli wisely refused, because he thought that, in *E. coli*, three chromosomes was a very unlikely proposition. He was right.

Meeting an American Genius (finally)

As we saw, Cavalli's work with the Lederbergs had never involved face-to-face communication, and had been communicated back and forth by airmail, including exchanging bacterial strains. When asked whether he found this situation difficult, Joshua Lederberg reported to us that "we [Cavalli and Lederberg] seemed to think along amazingly parallel tracks. We'd have letters crossing the Atlantic, explicating very similar results and interpretations."[2]

In 1954, Cavalli was finally able to meet his American colleague, Joshua Lederberg, thanks to a Rockefeller Fellowship which took him to Madison, Wisconsin, to collaborate with him. In our conversations, Cavalli used the word "genius" only twice (the first time, as we saw, to describe R.A. Fisher). The only other time he used the term was to qualify Joshua Lederberg (see fig. 2.8). Today, Lederberg is former president of Rockefeller University and is currently a leading adviser to the Pentagon on defense against biological warfare. Cavalli's admiration for Lederberg is reciprocated by the latter. Indeed, Joshua Lederberg told us in March 2003: "My regard, respect, and affection for Luca are unlimited."

Cavalli's stay in Madison would be of short duration, only three months. Nevertheless, there was enough time for him and Lederberg to do complicated experiments that demonstrated once and for all that all bacterial mutants appear spontaneously and are not the result of some kind of adaptation induced by the conditions under which they are cultivated in the laboratory. As usual with Cavalli, this study was highly quantitative and statistical.

Other efforts during that period were not so successful. As mentioned, the transfer of DNA between an *F+* or *Hfr* to an *F–* cell occurs through the formation of a kind of tunnel between the two cells, and this structure is

---

2. The correspondence between Cavalli and Lederberg can now be found archived on the Web site for the U.S. National Library of Medicine (*see* www.profiles.NLM.nih.gov/BB/Views/AlphaChron/series/02412/00267/01979). These letters make for a fascinating read and provide an original peek at the thinking of bacterial geneticists of the mid-twentieth century.

Figure 2.8 Cavalli and Lederberg (*right*) in Agrigento (Sicily, Italy, c. 1988). (From the collection of L. L. Cavalli-Sforza)

called a conjugation tube. In 1954 nobody had ever seen such a tube. Lederberg devised a very clever experiment to try to determine how DNA passed from one cell to another. To achieve this, he isolated an *F+* strain with round morphology that was nonmotile. In other words, this *F+* was unable to swim in liquid medium. Also, he isolated an *F–* that was elongated and motile—it could swim. When these *F+* and *F–* cells were mixed, allowed to mate, and observed under the microscope, the elongated, motile *F–* cells were seen pulling around the round, nonmotile *F+* cells. Clearly, some kind of structure was uniting and keeping together the two different types of cells. Lederberg attributed this phenomenon to cell surface "hairs"—called pili (sing., pilus)—bringing the mating cells in close proximity. Unfortunately, Cavalli and Lederberg were not able to visualize this structure; microscopic techniques were not advanced enough at that time. The conjugation tube, which allows the passage of DNA from cell to cell, was finally identified by others several years later (see fig. 2.9).

Cavalli was to return to Madison in 1958 to do some more work with Joshua Lederberg. Meanwhile, in 1957, another bombshell exploded in the world of bacterial genetics. Based largely on Cavalli's work, Wollman and Jacob of the Pasteur Institute in Paris (whom we met above), showed that chromosome transfer between *Hfr*s and *F–* cells took place gradually and

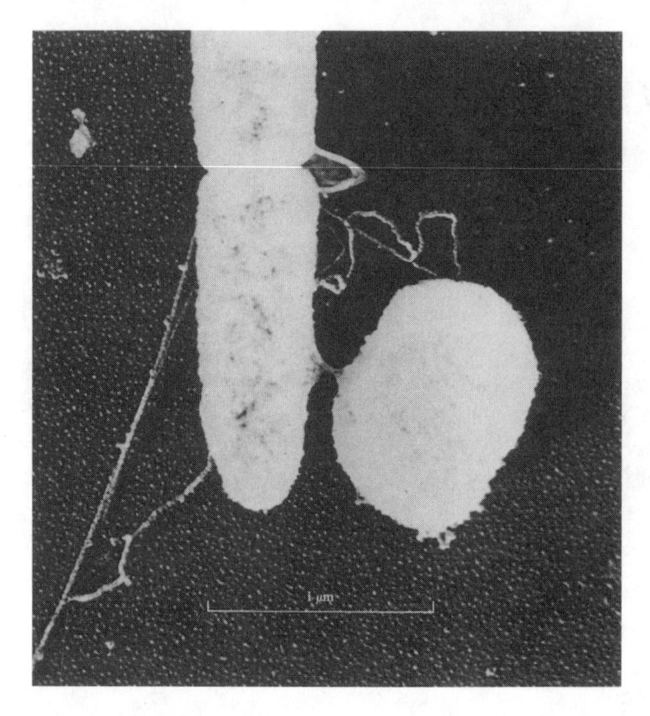

Figure 2.9 Electron micrograph of an *HfrF-* mating. The *Hfr* is the round cell and *F-* is the elongated cell. The bridge uniting them is the conjugation tube through which the *Hfr* chromosome is transferred. (From E. L. Wollman, F. Jacob, and W. Hayes. 1956. Conjugation and genetic recombination in *Escherichia coli* K-12. *Cold Spring Harbor Symposium on Quantitative Biology* 21:141. Reprinted with permission.)

slowly enough to time the whole process. This meant that some genes from an *Hfr* donor appeared before others in the *F–* cells, whereas others appeared later. In other words, gene transfer occurred in an organized, timewise fashion.

This discovery was met with enormous skepticism in the genetics community. Who were these two Frenchmen anyway? In 1958, Cavalli and Lederberg were not convinced and tried hard to disprove the French results, to no avail. The Frenchmen were right. From then on, the small world of bacterial genetics would swell at a very fast pace and become very crowded. Molecular genetics, with its cohort of complex and expensive equipment, would soon supplant the quiet labs where before only petri dishes, incubators, and water baths sat on benches. Bacterial genetics research in those early days did not involve machines, robots, radioactive techniques, and computers as it does today. For these pioneers, their brains were their most, and often only, valuable instrument.

Cavalli's Interest in Bacterial Genetics Slowly Wanes

In 1958, Italy, including the Istituto Sieroterapico in Milan, were still far from being the best places in the world to do highly competitive genetic work with *E. coli*. This was one of the reasons why Cavalli shifted out of the field of bacterial genetics. In addition, already back in 1952, as a part-time lecturer at the University of Parma, Cavalli had developed an interest in human genetics. As he once said, "humans are more charismatic than *E. coli*!" That much is certainly true. But then, a true geneticist never feels tied to a particular organism: he or she studies genes at large.

Cavalli's first foray into human genetics is notable. At Parma he found a student, Antonio Moroni, who was also a Catholic priest. As such, he had easy access to church records and so, under the direction of Cavalli, he started researching frequencies of **consanguineous** marriages (marriages between related persons—cousins, for example) in the Parma valley. Why consanguineous marriages? First, if a marriage is known to be consanguineous, one also knows at least part of a family tree over several generations. Then, marriages between consanguineous relatives restrict the genetic diversity that exists between the two members of the couple, as opposed to marriages between totally unrelated persons. And this is exactly what geneticists seek to do: observe offspring from parents with a known genetic background. Granted, determining the genetic background of humans is not the same as doing it with *E. coli*, where selecting strains for a particular trait is easy. Also, contrary to *E. coli*, people cannot be asked to mate according to the whim of the experimenter. Nevertheless, the genetics of human populations is not intractable and can indeed be very successful, as we shall see in the next chapters.

Having considered the pros and the cons of continuing work in the field of bacterial genetics, and having discovered a new way of combining genetics and statistics, by 1960 Cavalli's work with bacteria was over. He had become a full-time human population geneticist.

Chapter 3

The Shift to Human Populations (1952–1970)

The twenty years or so that followed World War II saw a period of turmoil for Italy. But this country, which had not been much known for its cinema industry before the 1950s, suddenly burst onto the international scene with names like Roberto Rossellini, Vittorio De Sica, Federico Fellini, Luchino Visconti (cousin of Cavalli's collaborator mentioned in chapter 2), and the greatest of the great, Michelangelo Antonioni (*L'Avventura*, 1960; *La Notte*, 1961). If you have not seen Italian neo-realism in movies, just rent *Roma, città aperta* (*Open City*, 1945), *Ladri di biciclette* (*The Bicycle Thief*, 1948), *Riso amaro* (*Bitter Rice*, 1949), and *La Strada* (*The Road*, 1954), and you will never look at black-and-white movies in the same way. In those days, Italy resembled some sort of massive social psychodrama, with writers like Italo Calvino, Curzio Malaparte, and Carlo Levi, and even pop idol Bobby Solo ("Una lacrima sul viso," A tear on the face), redefining society, liberating it from the long-lived demons of fascism, and bringing "the beat" to Italian popular music. Silvana Mangano was setting the standard for voluptuous and assertive Italian women, not yet Sophia Loren (who was at the time merely posing for nude pictures in her shabby Pozzuoli dwelling near Naples). The cinematic efforts of Italian directors would culminate in the late 1960s with the so-called spaghetti westerns, today seriously admired by film buffs around the world. Their creator, Sergio Leone, manufactured an American West of the mind, a West that could never have existed, but which was in line with the thinking of Leone's intellectual predecessors.

In politics, the initial euphoria that had accompanied the fall of fascism and the bringing to power of left-wing parties like the communists (the largest communist party outside the USSR was in Italy) and the socialists slowly waned and brought to power the ultraconservative Democrazia cristiana (Christian Democracy), also called "la DC" (pronounced "dee-chee"), the party that was to rule Italy for decades, helped by its underground "P-due" (P-two, pronounced "pea-doo-ay") secret organization composed of right-wing politicians, bankers, and otherwise rich and powerful men. Mercifully, this absolutely corrupt political party no longer exists today; it collapsed under its own scandalous weight in the 1990s. We will see later in this chapter that the "DC's" grip on political life contributed to the university student revolution in Italy in 1968.

Meanwhile, not much of this seems to have had an impact on Cavalli. He was absorbed in science and eager to engage in scientific research. As one of his colleagues put it to us many years later, "Cavalli is nearly glued to science." Clearly this was our impression as well. One could characterize Cavalli as politically moderate-to-liberal. For instance, while discussing his life in the United States, he praised the student protests against the war in Vietnam, he reported being "saddened" by the conservative Reagan years, and he was generally positive about the Clinton administration. When discussing the rise of feminism over this period, he said he was himself a feminist. Nevertheless, Cavalli seemed not to show much enthusiasm for any of our questions to him concerning politics or social issues; but when it came to discussing science, he lit up and became animated. He especially glowed when talking about his own experiments and their results, many of these having been conducted in the decades covered in this chapter. Before discussing what Cavalli did from the 1950s to the 1970s, it is necessary to explain in some detail his field of genetics.

The science of genetics is traditionally subdivided into three branches—classical genetics, molecular genetics, and population genetics. Oldest among them is *classical genetics*, which defines how genes are transmitted from parents to progeny, using the laws of heredity discovered by Mendel in the nineteenth century (see chapter 1). Classical genetics also deals with the mapping of genes on chromosomes, using techniques developed in the 1920s by American geneticists Thomas Hunt Morgan, Alfred Sturtevant, Calvin Bridges, and Herman Muller. Therefore, in terms of gene mapping, Cavalli's work on bacterial genes can be considered classical genetics, even though (as we saw) gene transmission occurs in a very peculiar way in bacteria. Classical genetics is also the branch of genetics that animal and plant breeders apply to develop new breeds of crop plants, ornamental plants, and domestic animals. Classical genetics can be studied and used without any knowledge whatsoever of the chemical nature of genes. In it, genes are simply points located at certain distances from one another on chromosomes.

Molecular genetics is a subfield that appeared in the mid- and late 1950s, and whose birth is traditionally associated with the discovery of the double helical structure of DNA in 1953. Molecular genetics deals, as one would expect, with the molecular mechanisms involved in the way genotypic instructions present in the base sequence of DNA are turned into the phenotypic properties of an organism, properties that largely depend on the nature of the proteins coded for by DNA. Contrary to classical geneticists and breeders, molecular geneticists use complicated and expensive equipment, and their science intersects deeply with chemistry and physics. *Population genetics*, which studies gene distributions in plant, animal, and human populations, is explained in greater detail in the next sections.

These three branches of genetics should not be considered independent from each other. Geneticists know very well that classical, molecular, and population genetics are three interwoven facets of the same reality—namely, the study of what genes are, how they work, how they are transmitted, and what their frequencies are (meaning frequency of occurrence) in populations. In fact, merging these three subdisciplines is seen as a very desirable goal in the case of human genetics, now that the human genome has been deciphered, and many disease genes and susceptibility-to-disease genes (including polygenes) await discovery. The distribution of these disease genes in the human population is of course of great importance.

The Science of Population Genetics

Population genetics has its roots in evolutionary biology. When Darwin published his famous book *On the Origin of Species* in 1859, Mendel had not yet discovered genes. Even though Mendel published his findings during Darwin's lifetime, there is no evidence that Darwin became aware of Mendel's explanation of the transmission of hereditary characters. Therefore, Darwin's original theory contains no genetics as we understand the field today. Population genetics can be seen as a unification of Darwin's and Mendel's theories, starting in the 1920s (with first attempts already made in 1908, however) and continuing today. Contrary to classical (or transmission) genetics, which concerns itself with how genes are passed from one *individual* to another, and molecular genetics which focuses on *cells* rather than individuals, population genetics focuses on genes and heredity in *groups* of individuals. In a sense, population genetics is the most holistic branch of genetics. It is also the most mathematical and statistical of the three subfields.

In a nutshell, population genetics seeks to understand genetic variation within and among populations and the causes for this variation. It is based both on measurements of gene frequencies as well as on mathematical models that try to explain changes in these gene frequencies and the natural forces that cause them.

Foundations of Population Genetics

Mendel's laws of genetics are statistical by nature. This is why genetic counselors, who evaluate concerned parents for the possibility of genetic disease in their offspring, can only give clients a *probability* of giving birth to an affected child. Further, when one studies populations, one generally deals with large numbers of individuals. Here again, one sees the usefulness of statistics for processing numbers.

One critical element of population genetics that follows is Mendel's demonstration that genes occur in nature as different variations on the same theme, or what we will here call variants (technically these are called *alleles*, short for allelomorphs, meaning alternative forms). This concept was already introduced in chapter 1, but it is so critical that it deserves repeating here with some examples. In humans, although the correct explanation is more complicated, one can think of a generic gene determining eye color, this gene occurring in the form of blue, green, brown, and so on variants. Similarly, the ABO blood group is under the control of a generic gene which exists in the form of three variants. In other words, to each generic gene there correspond several possible variants. When geneticists talk about and look at genetic variation in populations, this is what they mean and look for, the extent of variation in a population for a given gene. For example, a certain human population could show a given ratio between the *A*, *B*, and *O* variants, while another human population could show a different ratio. In technical terms, variations on a generic gene are called *polymorphisms*, and the gene itself is called a *polymorphic locus* (pl., loci) from the Greek *poly-* (many) and *morph-* (form) and the Latin *locus* (place). Many human genes have more than three variants. A further complication is that many genes can affect the same trait, independently or in unison. As a practical rule, differences in gene variants among individuals can be thought of as polymorphic if at least 1 percent of a population has one type of variant versus the other type(s). This is done to distinguish polymorphisms from rare mutations that are normally present at much lower frequencies.

But then, what exactly is a population for a geneticist? A population is a collection of individuals that are not only *sexually compatible* (they can mate because they belong to the same species, such as humans, flies, redwoods, etc.), but they also frequently *mate randomly*. Since most natural populations consist of organisms harboring different variants of many different genes, a population can be conceived of as a *gene pool.* Some will object that humans do not mate randomly and hence cannot be considered a gene pool. Whether one considers this prurient or not, human beings actually *do* mate randomly. Granted, we may select our mates for height, eye color, income, mental abilities, and other physical and psychological characteristics, many of which are at least partially under the control of genes. But we do not select mates based on what variants of immunoglobulin genes they have, nor on their blood group, nor on the variant of glucose-6-phosphate dehydrogenase or phosphoglycerate kinase gene they possess. In fact, we humans *do not* select mates for the enormous majority of the 25,000 genes we all carry.

Let us now move to a few very simple mathematical concepts that are at the core of population genetics. In 1908, G.H. Hardy, a Cambridge University mathematician, and Wilhelm Weinberg, a German physician, independently realized that the frequencies of gene-variant combinations (technically called *genotypic* frequencies) for any given gene in a population could be represented by a simple equation, called the Hardy-Weinberg theorem.[1] The equation can be easily adapted for cases where there are many variants, many polymorphisms, of a gene. The meaning of the Hardy-Weinberg theorem is quite straightforward: it can be demonstrated that, given certain conditions (which we will examine later), gene frequencies in populations do not change

---

[1] The Hardy-Weinberg equilibrium is written

$$p^2 + 2pq + q^2 = 1$$

where p is the frequency of one of the gene variants and q is the frequency of the other variant. By definition, $p + q = 1$ since the sum of both frequencies represents the totality of the variants of a single gene in a population. The following example illustrates the equilibrium. Let us assume a single human gene that possesses two variants, A and a. Since humans harbor two sets of chromosomes, some individuals will contain two copies of A (AA, the frequency of the homozygotes for A then being p^2, some will contain one copy of A and one copy of a (Aa, the heterozygotes, whose genotypic frequency is $2pq$), and others will contain two copies of a (aa, the frequency of the genotype homozygous for a thus being q^2). These are the genotypic variants. Whether one is AA, Aa, or aa can be determined by chemical analysis, and the numbers of genotypic variants in a population thus established. A population is in Hardy-Weinberg equilibrium if AA (p^2) + 2(Aa) ($2pq$) + aa (q^2) = 1. The equation can be used to calculate gene frequencies in two different populations—for example, the Basques and the Sardinians. Both populations have a common origin. However, among Basques, the frequency of blood group Rh-negative individuals (q^2) is 36% (or 0.36). Therefore, the frequency of the Rh-negative gene variant (q), the square root of q^2 , is 60% (or 0.6). Among Sardinians, q^2 is 4% (or 0.04), making q equal to 20% (or 0.2). This difference in Rh-negative gene frequencies (60% vs. 20%) might be due to natural selection of malaria resistance of the Rh-positive blood type, but drift might also have contributed in an unknown proportion.

with time, once the equilibrium predicted by the theorem is reached. That is, once the population is in equilibrium and mates randomly, gene and genotypic frequencies stay the same throughout all future generations. In part, the Hardy-Weinberg theorem was formulated for predicting the relative frequencies of genotypes of diploid species that mate randomly. From this, it is easy to calculate the frequencies of individual gene variants in a reproducing population. But then, this formulation is a mathematical model that may or may not conform to reality.

How plausible is it that human populations follow the Hardy-Weinberg theorem? It turns out that the theorem is extremely robust when applied to human populations. For example, various populations display different frequencies for the gene variants that determine blood groups, but these frequencies follow the Hardy-Weinberg equation. But also, blood group gene-variant frequencies, although they are in equilibrium in both Eskimo and Australian aborigines populations, show very different values. One can then wonder why this is the case. What are the factors that make gene frequencies different in different populations or, to put it differently, what is it that made the Hardy-Weinberg equilibrium different in Eskimos and Australian aborigines?

We know now that several factors can influence gene frequencies in populations. These factors are natural selection, migration, gene mutation, and genetic **drift**. For migration, consider an example of a human population where the frequency of gene-variant A is 100 percent and the frequency of gene-variant B is 0 percent. This population then migrates to a new area and mates with another population where the gene frequency for variant A is 0 percent and that of variant B is 100 percent. When the two populations mix and interbreed, the frequency of A in the mixed population will be somewhere between 0 and 100 percent, no longer one or the other, and the same holds true for B. Where exactly between 0 and 100 percent the new values will be situated depends on the relative numbers of individuals present in the two groups. Thus, migration, also often called **gene flow**, or **admixture**, changes gene frequencies.

Mutation, which has the ability to convert variant A into B—or vice versa—(and potentially change any gene), may or may not have an important effect on gene frequencies. As we saw in chapter 1, the simplest and most frequent mutation is the change of a DNA base pair (such as an A-T pair) into another base pair (such as a G-C pair). The magnitude of the effect of mutations on gene frequencies depends on mutation *rates*. Indeed, not all base pairs of human DNA, for example, mutate at the same rate. Low rates of mutation, as in most of our genes, have little effect on the equilibrium. But some segments of human DNA, such as microsatellites (to be defined later), mutate at higher rates. Gene frequencies for these DNA segments are modified accordingly. We will see that mutation rates sometimes facilitate the job of population geneticists and sometimes they hamper it, depending on the questions asked.

Then there is natural selection. As the name implies, nature can determine which variants of a particular gene and the phenotypes they determine are particularly well suited to a given environment. For example, a wing-color gene in some moths comes in two variants, dark and light. When dark moths rest on dark trees, birds that prey on them cannot see them easily. Many of the dark moths escape death and go on reproducing their dark variant gene. Conversely, light moths will be spotted quickly, eaten by the birds, and their light variant genes will be taken out of the gene pool. The reverse situation exists if the insects populate light trees, in which case it is the light variant that will survive and spread. This phenomenon has been observed in nature: where light trees turned dark because of industrial pollution, this led to the near-eradication of the light-colored moths that dwelled on them.

A similar situation also exists in humans. The genetic disease sickle-cell anemia occurs frequently only in areas of the world where malaria is prevalent. This is because the variants that cause the disease, through the making of abnormal hemoglobin, also provide significant protection against malaria. Thus, the defective variants are selected *for* by the parasite that causes malaria, and survive in the gene pool whereas the normal variants (which make normal hemoglobin) are selected *against* by the parasite. Therefore, the frequencies for hemoglobin gene variants in human populations differ according to whether malaria is prevalent or not.

Most of the time, except in some extreme cases (such as clearly detrimental genetic diseases), it is very difficult to decide whether human gene variants have definitive selective advantages or disadvantages. Nevertheless, natural selection can have a drastic and potentially rapid effect on gene frequencies. In many cases population geneticists studying human migrations (as Cavalli does), unless otherwise stated, use polymorphic loci thought to be selectively neutral. This simplifies their work because, in these cases, one can ignore the effects of natural selection on these loci. Work in progress at Stanford and elsewhere, however, is also attempting to determine which genes at large have been influenced by natural selection, which undoubtedly took place in humans.[2]

---

2. One example of a trait in humans which is definitely not selectively neutral is that of skin color. Dark pigmentation, owing to the pigment melanin, decreases the negative effect of ultraviolet (UV) radiation from the sun on the compound folic acid present in the skin. With high insolation, a light skin is detrimental because folic acid is destroyed, an effect that leads to abnormal births characterized by neural tube defects. However, some ultraviolet light absorption by the skin is indispensable for the synthesis of vitamin D. Under high insolation conditions, people with dark skin absorb enough UV to make adequate amounts of vitamin D. Under low insolation, the situation is reversed. There,

Finally, there is the effect of genetic drift. Drift applies especially when only a small segment of a population survives to reproduce or migrates elsewhere. This could happen after a cataclysmic event where most of the individuals in a population are wiped out, or it could happen if a small proportion of a population migrates elsewhere, splitting from the main group. These events are called **bottleneck** effects because the size of the population is severely decreased. To illustrate this, imagine that an original population of 10,000 humans consists of 70 percent individuals with an A variant of a gene and 30 percent of the B variant. Imagine further that two males and two females decide to leave the group and found a new population far away from the main group. It is unlikely that these four individuals will represent the original 70 percent A and 30 percent B variants present in the large population. It is in fact quite possible that all four of them would be A or all would be B. It is not even necessary to imagine that the bottleneck has to be of the magnitude specified above; time itself can deeply alter gene frequencies in all populations, more rapidly so the smaller these populations are. This is because only a *sample* of a population reproduces at each generation, thus creating a bottleneck.

Figure 3.1 shows a simulation of drift in such a hypothetical population. It can be seen that, purely by chance, the frequency of a gene variant, q, set originally at 0.5 (or 50 percent) can fluctuate and change drastically over the generations. In one case (a), the frequency of the gene becomes 100 percent after 32 generations. In another possible scenario (b), this frequency falls to zero after 30 generations. The principle at work here is called *sampling error*, meaning that if the number of progeny resulting from matings is small, this small sample may deviate from frequencies present in the larger pool. This is similar to flipping a coin. If we flip a coin many times, we expect heads or tails to come up equally often (i.e., 50 percent of the time each). However, if we flip the coin only four times, we would not be astonished to see all tails every time, or three tails and one head. The conclusion is that outmigration of small numbers of individuals or large population reduction can profoundly alter gene frequencies relative to the original population. On the other hand, unless a population is very small and considered over a short time frame, the

light-skinned people are at an advantage because dark skin would prevent the synthesis of vitamin D. Also, with low UV, the destruction of folic acid does not take place. Thus, as people moved out of Africa to occupy northern climes, light-skin variants were progressively selected for. On the other hand, early human populations that migrated out of Africa to tropical areas (such as the Melanesians and Australian aborigines) retained their dark skin pigmentation. Eskimos, and to a lesser extent Saami and other Arctic populations, have relatively darker skin even though they live at high latitudes because their food is naturally rich in vitamin D.

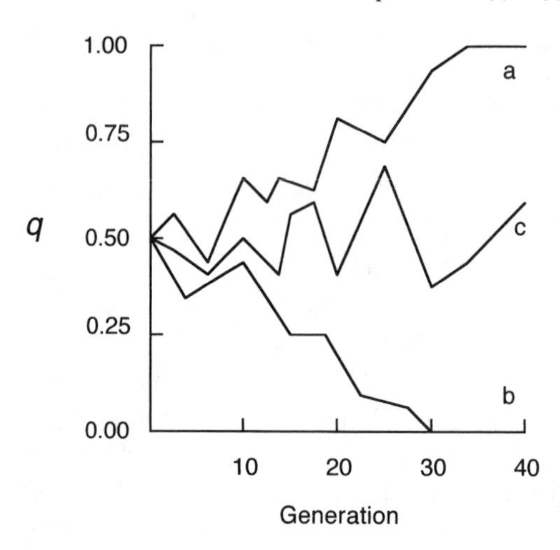

Figure 3.1 Simulated genetic drift in three populations. All populations start out with the same gene frequency q of 50 percent. Over time, drift causes fixation of that gene in population "a" and extinction of the same gene in population "b." In "c" the gene frequency varies widely over time but the gene neither disappears nor becomes fixed.

gene frequency can fluctuate significantly but not lead to the disappearance or fixation of the gene, as in (c) of figure 3.1.

Let us go back to the example of Eskimos and Australian aborigines that are both in equilibrium for blood groups, but with very different gene and genotypic frequencies. We can start to guess that perhaps some type of out-migration (if we ignore the effects of natural selection) from an original and ancestral human gene pool is responsible for this difference. Was genetic drift at work here? We will see that the scientific explanation is just that: small numbers of humans originally appeared a long time ago, split into groups, and became established in the four corners of the earth. Bottleneck effects were certainly—but not only—responsible for who we are genetically. We will see next that much of Cavalli's attention, at the beginning of his efforts in human genetics, was drawn to the study of drift and gene flow in human populations. His reasoning was that demographic data available in human populations would allow one to predict migration and drift effects that may account for human genetic diversity.

In summary, genetic divergence among populations, the study of which is at the core of Cavalli's research efforts, can be affected by mutation, which increases divergence (but only slowly over time) and gene flow, whose effects vary with population size and gene-variant frequencies in the mixed populations. Drift can have a significant effect on gene frequencies all the time, but in a marked way only after severe reduction of population size, whereas natural selection can increase genetic differences among populations by favoring different variants in different populations, given enough time.

Molecular Population Genetics

Today, population geneticists apply their equations to DNA itself. To do this, they isolate DNA from human cells and determine the sequences of a number of genes. Differences in DNA sequences among individuals can be thought of as polymorphisms, just like eye color and the ABO or Rh blood groups. These DNA polymorphisms can also be treated statistically and their distributions inside and among populations estimated. Later in this book, we will focus on the sequences of genes present on the Y chromosome and mitochondrial DNA. Non-Y chromosome genes will not be ignored, however, because much of Cavalli's work has focused on these genes.

The Y chromosome will be described in more detail in chapter 6. For now, suffice it to say that the Y chromosome is present in males only. Since it is exclusively transmitted from fathers to sons, because males carry one Y and one X chromosomes, the Y chromosome gives us information about the evolutionary history of human males. Mitochondrial DNA, on the other hand, is transmitted by women to both their sons and daughters. This is because sperm cells from the father do not contribute mitochondria to the fertilized egg. Mitochondria are small bodies, about as big as bacterial cells, that reside in the cell sap whereas chromosomes are present in the cell nucleus. Mitochondria metabolize the oxygen we breathe and, in that process, release energy which is used by cells to drive their metabolism. Mitochondria also possess DNA, which is distinct from the majority of the DNA present in chromosomes. This DNA is very small; it contains only about 16,600 base pairs (as opposed to 3.1 billion base pairs in chromosomal DNA). In spite of its small size, mitochondrial DNA is highly polymorphic either because it has a higher mutation rate than chromosomal DNA or through the effects of natural selection. Since it is transmitted through the mother, mitochondrial DNA can be used to trace the evolutionary history of human females. Starting in the 1980s, Cavalli used all three types of DNAs—non-Y chromosomal, Y-borne, and mitochondrial—to conduct his work on human variation.

Cavalli's Early Contributions to Human Population Genetics

While slowly abandoning bacterial genetics, Cavalli was laying the foundations for his later grand synthesis of human biological and cultural evolution. Already in 1952, while he was a part-time lecturer at the University of Parma, he had turned his attention to human genetics. His initial research would involve a single tool, the genealogies of the people inhabiting the Parma valley (rightly famous for its excellent parmigiano cheese and its ham, prosciutto crudo) in Northern Italy. No complicated laboratory equipment was required, just people's records of where they lived, their blood group, and whether they had married first or second cousins—in other words, whether they had engaged in consanguineous marriages.

The Parma valley consists of many villages and towns distributed both on flat land and in mountainous areas. This type of geographic distribution is in fact quite common in Italy. From parish books and from records of dispensations for consanguineous marriages granted by the Catholic Church, Cavalli made a careful study of the well-known fact that consanguineous marriages (leading to inbreeding) were much more frequent at high altitudes (measured from 0 to 1,100 m above the valley floor). Further, he observed that in the mountains, over a period of three hundred years, only 10 percent of the surnames were new. In the valley, by contrast, it took only sixty-eight years to witness the appearance of 50 percent new surnames. In addition, the average size of a mountain village was ten times smaller than that of a village in the plains. In other words, the sizes of population conglomerates and population movements between various places were definitely larger in the plains than they were in the mountains, with intermediate values in the hills. The idea of considering genetic drift (versus other evolutionary factors), that is, genetic effects following the isolation of some human groups, as an important tool in the study of the prehistory of human populations, germinated at that time in Cavalli's mind.

Drift as a Major Factor Responsible for Human Genetic Variation

Between 1951 and 1961, Cavalli lectured at the universities of Parma and Pavia. In 1961, Cavalli was offered a professorship in the genetics department of the University of Parma and then moved to the University of Pavia in 1963, where, the reader will remember, he had been a medical student. There, in collaboration with other scientists—including Antonio Moroni (who had been his student in Parma in 1951 and was to become a professor at the University of Parma) and Anthony Edwards, future Cambridge University professor—he developed sophisticated mathematical techniques to study drift and migrations in humans (see fig. 3.2). These studies were still restricted to the Parma valley. But, and this is important, this in-depth study of a rather narrow part of Italy allowed him to put his theoretical models to the test.

Figure 3.2 Cavalli, with Anthony Edwards (*left*) at a meeting. (From the collection of L. L. Cavalli-Sforza)

In science, a researcher develops a hypothesis, an idea, a model, to explain some observed phenomenon. Then the researcher must test this hypothesis by making additional empirical observations. If observations concur with the model, this model can then be expanded, made more complicated, and, at a certain point, this model evolves into a theory. Thus, a theory is much broader than a single hypothesis, and it addresses more general questions. Cavalli's thought process followed this path; he first developed a simple model based on simple observations in the Parma valley. Subsequently, he formulated a much more complex theory of human migrations on a world scale.

Cavalli's earlier hunch that drift might be at work in the Parma valley was supported in 1963, when he measured the variations in three blood systems as a function of village altitude (and population density). The three systems he used were the *ABO*, *MN*, and *Rhesus* (*Rh*) polymorphic genes. His observations on blood group heterogeneity were consistent with drift being at the root of this heterogeneity, owing to isolation and low population density. At the time, in 1955, these three polymorphic systems were just about all that existed for genetics research, and they were cheap and easy to determine. In the next decades, Cavalli would considerably extend this number of genes and apply the study of human variation to the whole world.

Cavalli's work on genetic variation in the Parma valley began in 1955 and lasted until 1963. In order to determine whether drift or natural selection was responsible for the observed genetic variation, Cavalli and coworkers made use of a new instrument: the computer. Dissatisfied with the rigid mathematical theories that existed at the time, Cavalli realized that computer simulations might prove more flexible in the study of human populations and genes that vary so much in numbers and migration rates. The University of Pavia procured its first computer in 1963. It was awkwardly large (as they all were in those days) and had a memory about one millionfold smaller than a modern desktop computer. Cavalli and his associates were the only ones to use it, and it proved invaluable in their hands.

Moroni, being a Catholic priest (as we earlier explained), had access to parish books of births, marriages, and deaths. Therefore, he, Cavalli, and their coworkers had access to a nearly complete demography of the Parma valley, which they used in their computer simulations. Briefly, they simulated a population of 5,000 individuals from the twenty villages at the highest altitude, where drift, as they had inferred, was the greatest. Every simulated individual was born, married, procreated, migrated, and died, just like the real people recorded in the parish books. All simulated individuals were equipped with the three Mendelian genes (*ABO*, *MN*, and *Rh*) mentioned above. This study, according to Cavalli, was certainly the first simulation of the genetics of a human population. Their results showed that genetic drift best explained the actual gene frequencies found among the inhabitants of the high hills. This did not mean that natural selection did not exist; it simply meant that in the populations they studied, it was not strong enough to be observed.

In 1968, Cavalli's good friend, the great population geneticist Motoo Kimura, published a revolutionary article that generalized the concept that drift, and not just natural selection, plays a very important role in molecular evolution. As Cavalli puts it: "It [Kimura's article] clearly was of much more general importance than my discovery, covering not just a short stretch of time and space as mine did, but extending the same conclusion to nothing less than all of evolution. But in the end, it was the same."

At this point in our story it is important to give a brief history of the evolution of ideas in mathematical genetics. Drift as an evolutionary factor had been considered earlier by others. But as neo-Darwinism was being developed in the 1920s, all mathematical geneticists focused their attention on evolution by natural selection. Their thinking was that drift alone could not maintain organisms alive, only natural selection could. Then, in 1941, Sewall Wright first proposed that drift was an important factor in evolution. His ideas were further developed by the French scientist Gustave Malécot, whose work was subsequently disseminated by J. B. S. Haldane, Ronald Fisher, and Sewall

Wright. However, this first generation of mathematical geneticists, who operated between about 1920 and 1950, had no knowledge of the precise molecular nature and consequences of mutations.

Later, as molecular genetics was developing, it became clear that fresh ideas, based on molecular data, had become necessary. Motoo Kimura, with his neutral theory of evolution, was the first to realize that most mutations are selectively neutral, leaving much more room for drift to play an important role in evolution. What is more, in the absence of strong positive or negative selection for most mutations, chance alone in the form of random mutations can cause genetic differentiation in many functionally meaningful directions. Thus, most mutations are not under the control of environmental factors—which by definition constitute natural selection—and are not automatically either eliminated or favored. In that sense, deleterious or advantageous mutations tell us more about the history of the environmental challenges that organisms have had to meet over time. Thus, neutral mutations under the control of drift are more useful in tracing the history of organisms themselves. Let us now go back to Cavalli's work in the 1960s.

In the 1960s, Cavalli developed models for human migrations, using again the Parma valley as an experimental ground to test his hypotheses. To do this, he studied matrimonial migrations (i.e., who moves where to get married), whether brides migrate more often than do grooms, and what kinds of parent-offspring migrations take place. Then in 1963–64 he established with Anthony Edwards a set of mathematical techniques, a theory, which allowed them to tackle the general problem of the history of human genetic diversification at the world level. The level of complexity of the theory, together with massive data collection (one of his samples consisted of 15,647 consanguineous marriages, later to become over 500,000), made it necessary to use computers again, which in those days (as noted) were orders of magnitude slower than current computers. It is also at that time that Cavalli made the first use of *principal component analysis* and phylogenetic trees to study human diversification.

Before discussing **principal components** and trees of descent, we will first introduce the useful concept of *genetic distance*. Genetic distance is not a difficult concept. Let us assume two different populations. If the first population shows a 30 percent (0.3) frequency for a given gene, and the other population shows 20 percent (0.2) frequency for that same gene, the genetic distance between these two populations for that gene can be taken as equal to the square of the difference between their gene frequencies, that is, $(0.3–0.2)^2 = 0.01$. One can perform this simple type of calculation for many different populations compared pairwise. Once this is done for many genes, one calculates averages that give a proper weight to each gene.

One can see that in order to study human evolution (or plant evolution, for that matter), one must perforce categorize individuals, putting them into genetic classes or categories. According to Cavalli, this difficult part of the research should be given more attention. For example, one can take a given population and determine how many individuals are Rh-positive and how many are Rh-negative. Continuing to do this for many other genes, one ends up with a genetic profile of a population. One can do the same thing for many different populations scattered around the world. Then, using the sophisticated mathematical models established by Cavalli and coworkers, one can determine to what degree different populations are genetically related, and one can calculate their time of evolutionary separation. To do this, one first uses a formula which allows the calculation of the variation of gene frequencies among the studied populations. This quantity is called F_{ST}.[3] Its formulation was published by Sewall Wright in 1943.

The value of F_{ST} varies between 0 and 1. A zero value means that there is no genetic variation among the observed populations. This is equivalent to saying that all the genetic diversity within a species is shared equally by all populations. A value of 1 means that all the genetic diversity within a species is observed among different populations, with no genetic diversity within these populations. For the human species $F_{ST} = 0.16$, which is the genetic distance between two populations averaged for many genes. A value of 0.16 is low, especially considering that this number is obtained by tallying polymorphic genes, which by definition differ from individual to individual. This means that in humans, only 16 percent of the total genetic variation corresponds to genetic differences *among* populations or groups. This also means that 84 percent of the human genetic diversity is observed *within* populations. For comparison, F_{ST} is 0.29 for coyotes, 0.39 for African elephants, and a high 0.85 for gray wolves. This means that, for gray wolves, the genetic diversity within a pack is only 15 percent, indicating that all wolves within a pack are genetically rather homogeneous. Conversely, 85 percent of the genetic diversity in gray wolves can be attributed to differences among packs. Different packs are thus genetically quite heterogeneous when compared with each other. This situation is reversed in humans.

[3] The equation defining genetic variation is written

$$F_{ST} = V_P / p\,(1-p)$$

where V_P is the variance (variation) between gene frequencies of any number of populations, and p is their average gene frequency. Both V_P and p can be measured in the field by using protein or DNA polymorphisms.

Then, one uses the value of F_{ST} in another equation, which correlates it with evolutionary time measured in human generations, and population size.[4] F_{ST} is but one measure of genetic distance. Many other such measurements have been proposed. At the beginning of their research on phylogenetic trees, in 1962 (with results published in 1964), Cavalli and Edwards designed a new way to measure genetic distance, based on a suggestion by R.A. Fisher. Another population geneticist, Masatoshi Nei, proposed in 1972 an alternative measurement that became rather popular. However, Nei admitted later that Fisher's approach, slightly modified, applied better to human populations.

When F_{ST} (or similar) values are plotted worldwide versus the geographic distances separating groups of individuals, the graph clearly shows that F_{ST} increases with distance. In other words, the greater the geographic distance between two populations, the greater their genetic distance, meaning the greater the difference between their gene frequencies. These differences must then be interpreted in terms of what phenomena caused them to take place as a function of space (and presumably time), such as, for example, drift and natural selection.

Estimating genetic distances automatically means categorizing people, putting them in little boxes of sorts. Some people carrying a certain type of genetic makeup end up in box number 1 while those carrying a different gene combination end up in box number 2, and so forth. There simply is no other way a human geneticist can go about studying the diversification of humans than by first classifying them into groups. And in reverse, building a descent tree for humans means acknowledging some genetic differences among them. However, categorizing people, putting them into classes that have any biological referent, can suggest a "racial" classification and then, potentially, racism. We all know that Negroes in America were categorized according to their skin color, and Jews in Nazi Germany were categorized according to their name and synagogue records.

---

[4] Genetic distance between two populations is defined as

$$D = -log\,(1-F_{ST})$$

where *log* is the natural logarithm. It is important that genetic distance show a simple relationship with the joint effects of migration and genetic drift. For human populations, it can be shown that F_{ST} approximates $1/(4Nm)$ for **autosomal** genes and $1/(Nm)$ for Y chromosome genes and mitochondrial DNA genes, where N is the effective population size (number of parents per generation) and m is the fraction of parents that come from the outside. Drift is measured by Nm, because the greater the gene flow (m), and the greater the population (N), the smaller the drift. This is just one definition of genetic distance among several others.

Cavalli has made it very clear in his abundant writings that the notion of "race" is unscientific. We will go back to these issues later in the book. For now, suffice it to say that, from a long time ago, a person like Darwin understood that race is *not* a scientific concept. He himself explored the question and discovered that, according to the authors of his days, there were anywhere between two and sixty-three human races. Clearly, Darwin was telling us that discussing races was not useful; people could not agree as to what they were anyway. When pushed, Cavalli, who does not like the term *race*, recognizes that there may be thousands of different human populations characterized by particular gene combinations. And he means this in a *medical* sense, not in any kind of racial, social, or other context: that is, some groups are useful to distinguish for medical purposes. Nevertheless, Cavalli has had to confront accusations of racism (as we will see in chapter 7).

Things Get More Complicated

We once asked Cavalli why he had not tried to explain what statisticians call **principal component analysis** in his trade book *Genes, Peoples, and Languages*. Using scratch paper, we went over a brief review of the concept one day in Palo Alto, California, while having lunch. He then said, "*You* try to explain that in a simple manner!" Certainly, explaining the fine details of principal component analysis is beyond the scope of this book; nevertheless, we provide below a simple example that will help the reader understand what this methodology is about. Thus, here are some of the mathematical tools used by Cavalli and coworkers.

First, principal component analysis. This phrase refers to a type of complex statistical procedure called *multivariate analysis*. We saw that Cavalli and coworkers used hundreds of gene frequencies to track down human origins and diversification. The word *multivariate* simply covers the notion that multiple elements—the gene frequencies—all vary in different populations. In other words, gene frequencies vary in what can be conceived of as a multidimensional gene frequency space. Even though we can manipulate space with 4, 5, 6, etc. dimensions mathematically on a piece of paper, we, being three-dimensional creatures, cannot visually represent and understand spaces with more than three dimensions. This presents a serious problem if, for example, one wants to represent multiple gene frequencies on a map of the world.

Principal component analysis comes to the rescue and allows just that. But this comes with a price: the fragmentation of information. Basically, principal components come attached with a number—first, second, and so on. By definition, the first principal component extracts the most information regarding gene variation from the total information available. The second principal component extracts a little less information (information which is *not* explained by the first principal component, however), the third less still (but information *not* explained by the first and second principal components), and so forth, until 100 percent of the total information has been extracted from the data. In practice, one rarely goes beyond the seventh principal component because, at that stage, most of the recoverable information needed has been gathered, unless the phenomenon is particularly complex.

Figure 3.3 shows how the first principal component is extracted from a set of data corresponding to two gene frequencies in five populations. Here, the frequencies of the *O* blood group gene are plotted against the frequencies of the *Rh–* gene found in America, Australia, Africa, Europe, and Asia. The figure shows that the five data points are scattered in the plane of the graph. However, the relative positions of the data points can be simplified by projecting them onto a diagonal line drawn in such a way that the distance between the line and the data points is minimized (this line is thus different from a regression line). The data points are then projected at right angles onto the diagonal, thus reducing the number of dimensions from two to one. This is the first principal component. This projection process decreases the amount of information available from the two-dimensional graph (the scatter has been eliminated), but it does so in a measurable way. The first principal component extracted from this example shows that American and Australian populations are clustered, whereas one finds an African-Asian cluster at a certain distance from the first one. Europeans are at a greater distance from the Americans–Australians than they are from the Africans–Asians. The process can be repeated as more gene frequencies are added. For example, three gene frequencies allow the extraction of yet another principal component, the second principal component, and so forth. For simplicity, figure 3.3 shows the frequencies of only two genes. It is therefore very approximate. The simplification introduced by the use of principal component analysis becomes more and more precise as more genes are used for the analysis.

To illustrate the complexity of the questions that Cavalli and his coworkers addressed, and to understand why principal component analysis was necessary, here are a few numbers used in their research: they tabulated a total of 76,676 gene frequencies, corresponding to 120 different genes, from 6,633 different geographic locations corresponding to 1,915 different population names. This is a massive amount of data. If one wanted to represent in one single table all the individual gene frequencies studied by Cavalli, one would end up with a rectangle containing 120 gene entries multiplied by 1,915 populations, for a total of 229,800 bins. Such a table is not only extremely awkward to compose, it is also totally impossible for the human brain to visualize its meaning. On the other hand, principal components as used to summarize all the data allow *global* representation of gene frequencies.

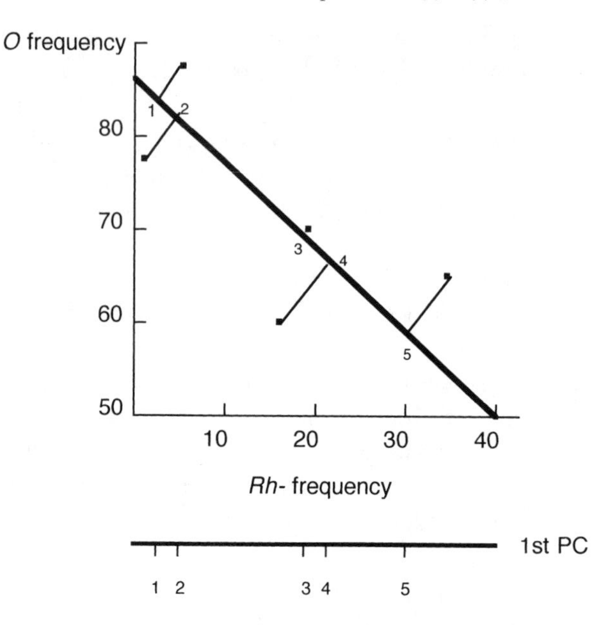

Figure 3.3 Derivation of the first principal component corresponding to two gene frequencies (*O* and *Rh–*) among five populations. *Top*: a two-dimensional representation of the correlations between the *O* and *Rh–* frequencies among Americans (1), Australians (2), Africans (3), Asians (4), and Europeans (5). The thin (right-angle) lines represent the projections of the data points on the diagonal (thick line). *Bottom*: first principal component (1st PC) reducing the two-dimensional graph to a unidimensional line. (Redrawn from L. L. Cavalli-Sforza, P. Menozzi, and A. Piazza. 1994.)

Figure 3.4 gives examples of the first and second principal components superimposed over a map of the world. The different shadings represent areas with different gene frequencies. The first principal component shows two poles, one situated in Africa and the other one located in Australia. Here, the first principal component (based on genetic distance, F_{ST}) represents nearly 35 percent of the total genetic variation observed in the world. This map represents a split between Africans and non-Africans. Since genetic distance measured by F_{ST} is approximately proportional to time of separation, we will see later that this observation is consistent with the idea that African populations started differentiating about twice earlier than the rest of the world did and that, therefore, Africa accumulated more internal genetic heterogeneity than all the other continents. This observation is also consistent with the idea of a small subset of an African population migrating to Asia, with subsequent spreading to the rest of the world.

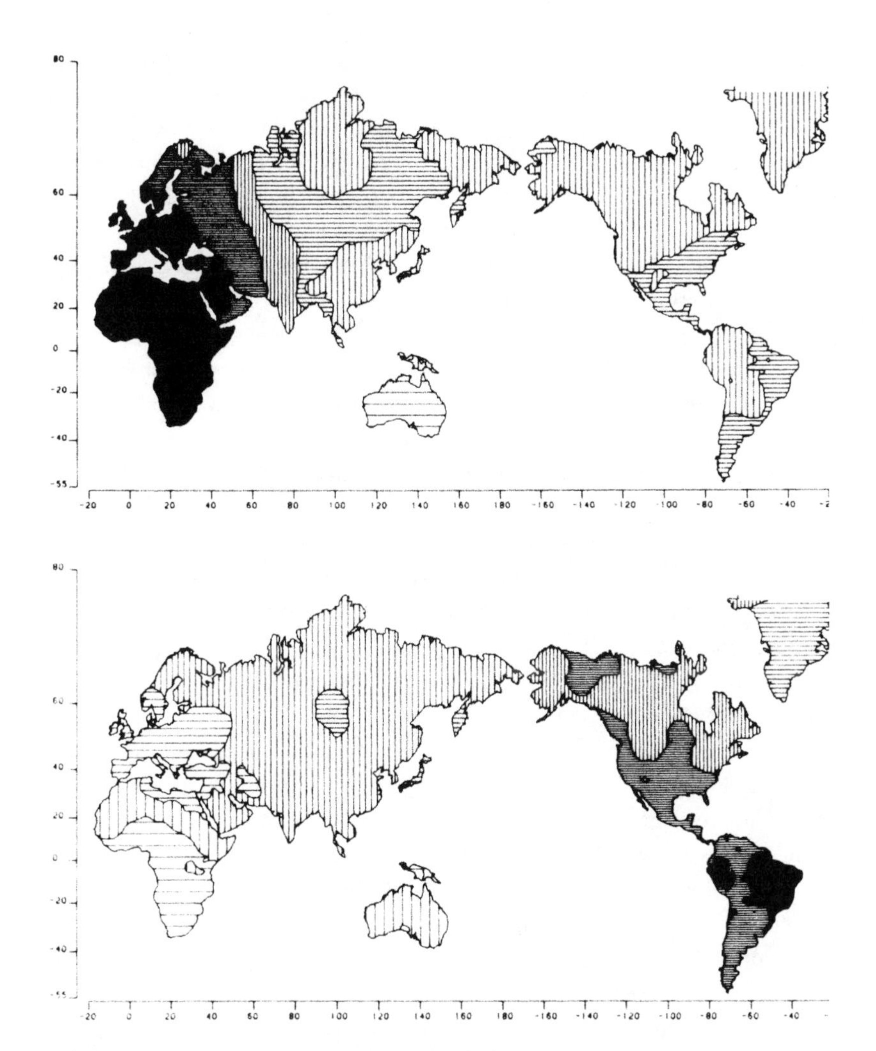

Figure 3.4 First and second principal component gene maps of the world. *Top*: map based on the first principal component. *Bottom*: map based on the second principal component. In both maps the range between the maximum and minimum values of the principal components has been divided into eight classes. (From L. L. Cavalli-Sforza, P. Menozzi, and A. Piazza. 1994. Reproduced with permission)

The second principal component represents the next 18 percent of the total human genetic variation. It displays one pole in Australia and the other one in the Americas, while Asia and Europe are in the middle of the axis of the second principal component. This map is thought to represent a further split in the human family, one separating Australia and Southeast Asia from the rest of the non-Africans. Thus, principal component analysis represents gene variation in space and, with the help of F_{ST} that can be correlated with time, variation corresponding to evolutionary distance. Moreover, combined with other information (such as archaeological data), principal components can be seen as population splits resulting in the settlements of new territories over time. For example, the third principal component of the world shows that European populations expanded from Asian populations.

At first sight, principal components may not say much about the history of human genes—that is, their diversification over time, which is best studied by building phylogenetic trees. But in fact, Cavalli and colleagues demonstrated that the two methods are strongly related. For example, they showed that the order of the splits found in phylogenetic trees in the absence of cross-migration between neighbors, and assuming independent evolution in the various branches of the tree, corresponds in fact to the order of principal components, first, second, third, and so on.

We will return to principal component maps and their meaning in chapter 5. It should be realized that building such maps was no simple matter. Cavalli and his coworkers spent many years inventing and refining statistical and computer techniques to arrive at results that are easy to visualize. However, the group of Robert Sokal, a statistician at the State University of New York–Stony Brook, questioned the validity of principal component maps in a 1999 article. An arcane discussion on the statistical accuracy of these maps ensued, but had no bearing on their overall biological meaning. In fact, Sokal and Cavalli agree on the genetic data that support population movements into Europe from the Middle East during the **Neolithic** period, as first shown by Cavalli and associates (see chapter 4). This was confirmed years later by Sokal (in spite of Sokal's ill-inspired—and later vaguely retracted—claim to precedence in this matter). It remains that, in spite of their visual attractiveness and methodological innovation, geographic maps of individual principal components are less used today mainly because they are not needed to represent gene frequencies in mitochondrial and Y-chromosome DNA now used to trace human population movements (see chapter 6). However, two-dimensional displays showing populations as points in diagrams representing the first two principal components remain popular.

Phylogenetic trees, or trees of descent, represent genetic variation in time and thus provide additional information adding to the geographic distribution of genes we just reviewed. Cavalli and Edwards first pioneered this type of analysis with human populations on a world scale in 1964. We have seen examples of such trees in chapter 1. As discussed there, trees can be constructed linking humans to their ancestors, or linking them to modern primates. In these trees, genetic distance between populations is combined with time (expressed in years or human generations) to estimate at what point in the past two populations separated and developed different gene frequencies. Figure 3.5 gives a theoretical example of the concept. In this figure, genetic divergence is represented as a function of time. At the top of the diagram, for example, one finds a single ancestral population that splits into two subpopulations at a certain point in time through drift. Later, each of the two branches splits again and produces a total of four sub-subpopulations. These are populations 1, 2, 3, and 4. These populations are present in the "TODAY" plane, which is the one we observe directly. These four populations may also be present in different geographical locations. The "TODAY" plane shows variation between the four populations for only two genes. For example, population 3 has a very high frequency for the A gene and a medium one for the B gene. Conversely, population 4 has a rather high frequency of A, but a low B frequency.

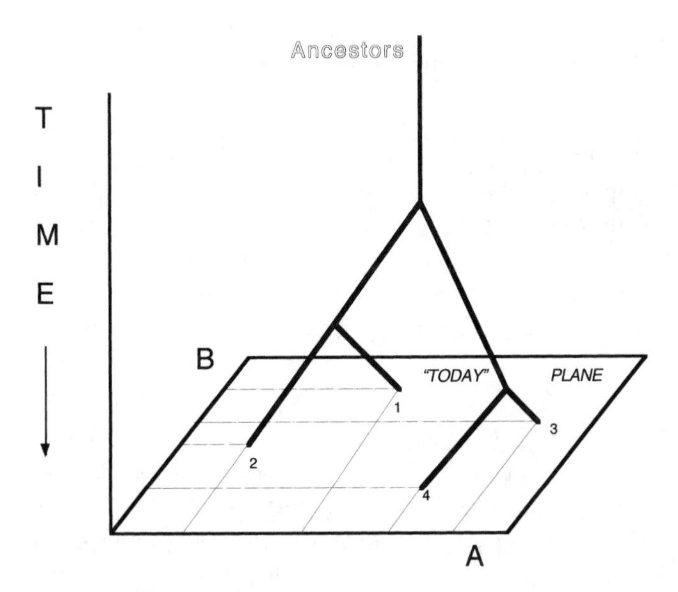

Figure 3.5 Differentiation of a single ancestral population into four contemporary populations. Drift causes genes to appear at different frequencies in the four populations. (Redrawn from L. L. Cavalli-Sforza, I. Barrai, and A. W. F. Edwards. 1964.)

One realizes immediately that this type of representation, useful as it is to understand the concept of population divergence, is impractical when more than two gene frequencies are considered. For fifty genes, we would have to imagine trees in a 50-dimensional space, which is an impossibility. Therefore, phylogenetic trees in Cavalli's work are represented exactly as in chapter 1, with the understanding that these trees represent the sum total of the genetic variation found between populations. Here again, it took Cavalli and coworkers many years of theoretical work to establish the complex statistical and mathematical foundations that led to both a geographic and a historical picture of human descent and diversification.

Finally, Cavalli and Edwards established another statistical technique called *cluster analysis*. As discussed, population genetics involves grouping organisms into categories—clusters—to determine their similarities and differences. In a 1965 article that some would consider mind-bending, Cavalli and Edwards show that their cluster analysis is particularly well designed for use with computers and crunching of large numbers. Basically, their technique applies to any set of objects, or points, that must be arranged in multidimensional space. They also demonstrate that cluster analysis and "tree" building are mathematically related. By any account, such intellectual efforts firmly deserve the adjective "academic."

But during these years, Cavalli was not just hunched over his theoretical equations. He also traveled to very unusual destinations to perform fieldwork in anthropology and genetics. In particular he started visting at this time a number of Pygmy populations in Central Africa. Cavalli was interested in these populations both genetically and culturally. In terms of his anthropological interest in how culture evolves, he was drawn to these societies as representing the hunter-gatherer mode of subsistence, which is similar to the general human hunter-gatherer adaptation in the **Paleolithic**.

Encounters with African Hunter-Gatherers

In some ways, Cavalli is a daring adventurer. Once he crossed the Sahara Desert (with his wife, Alba) in a Land Rover. This, he said, was the most exciting experience of his entire life. He went on to describe the continuously changing, beautiful landscapes of the desert, its challenge and dangers, and the untouched nature of the environment. He also told us that in 2003 he suffered heat exhaustion in the East African deserts, in spite of his strong physical resilience, while in search of rare genes. Cavalli must also have faced many a peril in the form of green and black mambas (the most poisonous snakes known to humans, sometimes called "minute snakes" by natives for a reason that is easy to understand), sleeping-sickness-carrying tsetse flies, and leopards in the jungles of Central Africa. After reading this last sentence, completely crafted by us in our manuscript, Cavalli commented to us that all these dangers have a low probability of occurrence, as they are "more or less like winning a big prize at a lottery, an unlikely occurrence."

Between 1966 and 1985, Cavalli mounted ten expeditions to study many groups of the very interesting peoples collectively known as Pygmies in Central Africa. There are a number of different Pygmy populations in different African countries. Cavalli worked mostly with the Aka of the Central African Republic and Cameroon, but he also visited several other groups, such as the Mbuti and the Baka of northeastern Congo. Cavalli was interested in studying genetic differences between Pygmies and other African peoples. Pygmies are, for one thing, distinguishable by their short stature; in smaller tribes the average height of men is four feet nine inches. This is an adaptation to their hot tropical forest environment where small stature is an advantage. Small body size helps to keep the body cool and uses less energy. A second reason why Cavalli wanted to study these people is because they are active hunter-gatherers (few other people in the world still rely primarily on hunting and gathering for subsistence). Over 99 percent of humans' time on earth has been spent in the hunting-gathering mode of adaptation. Cavalli saw that understanding this adaptation better can help us to understand aspects of human evolution.

In Cavalli's line of work, however, studying genes means taking blood samples. This proved to be not so easy. One time in Africa, he traveled with the great Scottish anthropologist Colin Turnbull, while trying to collect blood samples from some Mbuti school children (see fig. 3.6). Here he was confronted by an axe-carrying father who told him, "If you take blood from my son, I'll take yours." Hearing this, Turnbull dryly commented, "Perhaps we should leave now . . . " Initially, Cavalli also encountered some resistance to his taking blood samples among the Aka. His mistake was to work though an intermediary, a French coffee plantation owner. Cavalli wrote:

> I presented myself on the appointed day, with my companions and equipment, only to find that all the pygmies had disappeared into the forest. Word had spread that I was a *likundu* (a demon or bad witch). They left their village idiot behind for my tests—I never learned whether this was done in scorn or to see what I would do to him. (Cavalli-Sforza and Cavalli-Sforza 1995:5)

From then on he avoided intermediaries with the Aka, and it worked. "Taking blood samples from the Pygmies became extremely easy, much easier than from any other people I worked with before or since" (Cavalli-Sforza

Figure 3.6 Cavalli with Colin Turnbull (*far right*) and Turnbull's student in Austria. (From the collection of L. L. Cavalli-Sforza)

and Cavalli-Sforza 1995:5). What may have helped was Cavalli's idea to bring along presents in the form of soap (which the women preferred but the men largely ignored), salt (sodium is a highly desired and rare commodity in the jungle), and, yes, the formerly ubiquitous gesture of friendship: cigarettes. In later expeditions, Cavalli skipped the cigarettes, however.

Cavalli wanted to study the culture of these tropical forest hunter-gathers because, as he told us, "I was flabbergasted that we could be so different." It was clear during our interviews that Cavalli is genuinely fond of and admires these successful hunter-gatherers. He also once commented, "I wish I had a Pygmy's blood cholesterol level." In 1986 he published a book about Pygmy populations. Once he wrote: "It was a considerable surprise, to me and to all the co-workers who have been in the field, to discover that these most 'primitive' people are among the most charming, kind, and socially mature people whom we have come across." He continued: "Pygmies are extremely brave hunters and still, perhaps, the least aggressive people on earth. They have learnt to solve their problems of interhuman contact by techniques which we seem to have largely forgotten." (Cavalli-Sforza 1971:94).

How can one characterize the culture of these forest hunter-gatherers, and in what sense does their culture teach us something about human evolution?

The cultures of these populations share many characteristics with that of other hunter-gatherers. These include a very fine-tuned adaptation to the environment and a way of life that, contrary to that of industrial nations, preserves rather than destroys the environment. Hunter-gatherers are also known for their informal leadership and egalitarian social relationships. Anthropologists differ in the extent to which they perceive true gender equality in hunter-gatherer societies, but most agree that in these groups men and women are more equal in power, status, and access to resources than in agricultural societies. Hunter-gatherer groups have worked out a fulfilling and satisfying way of life that appears to produce little stress or conflict. The Aka, Mbuti, and other similar populations have been particularly noted for their peaceful and gentle way of life, their bravery and skill at hunting, and their ability to extract medicines, poisons (used in hunting), and food from the tropical forest. According to anthropologist Barry Hewlett, along with relatively high gender equality, Aka men spend a great deal more time caring for and raising children than do men elsewhere.

Another characteristic that these people share with other hunter-gatherer groups is the threat of cultural extinction they now face. For over 100,000 years all humans on earth lived as hunter-gatherers; now less than 500,000 of the world's 6 billion people do so. The development of agriculture, Western expansion and colonization, industrialization, and now global capitalism leave a grim tale of exploitation, disease, and loss of resources for indigenous peoples the world over. These peoples, aptly termed by anthropologist John Bodley as "victims of progress," have everywhere lost their land to outsiders and are given little choice by national governments but to become disadvantaged minorities in a new socioeconomic order. What happened to Native Americans is now happening to Pygmies as neighboring populations clear forest land for farming and extract timber.

In fact, it was his first expedition to Central Africa that spurred Cavalli's interest in cultural evolution (see fig. 3.7). "I started trying to understand something about cultural evolution," he said, "when I saw how different the Pygmy way of life was from that of others."

Cavalli also found that the Aka, for example, have experienced little genetic or cultural flow from their neighbors, the Bantus except in the case of the Biaka (West) group. Today, Aka genes flow to Bantus, because, occasionally, Bantu farmers marry Aka women. The opposite is much more rare.

Back in 1971, Cavalli was formulating his theory of **demic** (people's) **diffusion** to explain how agriculture was brought to Europe by migrating Neolithic farmers from the Middle East. (This theory will be discussed in detail in chapter 4.) Interestingly, Cavalli wondered why the Aka (like other Pygmies) never became acculturated and did not interbreed much with, and never adopted the farming lifestyle of, the Bantus, which started expanding from their area

Figure 3.7 Doing field work with Pygmies (Central African Republic, ca. 1966–69). (From the collection of L. L. Cavalli-Sforza)

of origin in Cameroon toward Central and South Africa about three thousand years ago. Obviously, in Europe, hunting and gathering practices disappeared a very long time ago. Why, then, did the Aka resist cultural absorption when the Paleolithic populations of Europe did not? According to Cavalli, two factors may explain these differences. First, European winters are cold and not conducive to hunting and gathering. Farming, on the other hand, allowed production of storable food and relative independence from climate. These ecological conditions do not prevail in tropical and equatorial Africa where hunting and gathering can take place year round.

Then, sub-Saharan agriculture has always been problematic and continues to be so today. Problems with plant diseases and pests continue to plague farmers there. Since, undoubtedly, crops of several thousand years ago were inferior to present ones, agriculture may not have been seen as too attractive a notion to hunter-gatherers so well adapted to the jungle. Thus, these people never adopted agriculture because, contrary to Paleolithic Europeans, they did not need it. Unfortunately, as the tropical forests are gradually disappearing, so are the Central African hunter-gatherers.

However, the hunter-gatherers were not the only ones who attracted Cavalli's attention in the 1960s. Stanford University was also on his mind, because his old acquaintance Joshua Lederberg had moved there, and also because a man named

Walter Bodmer was working there as a professor of genetics. Walter Bodmer was born in Germany, in the city of Frankfurt am Main which Cavalli had coincidentally visited during World War II. At the time of this visit, Bodmer was six years old and had already emigrated to England with his family. He became a British citizen and went on to become Sir Walter and a very famous scientist in his adopted land. Sir Walter's early infancy had been deeply disrupted by events unfolding in Nazi Germany. His father was a Jewish medical doctor (his mother was a Gentile) with aspirations to academia. However, years before the Nazis took power in Germany, during the period known as the Weimar Republic, Bodmer's father had already been told that his hopes of becoming a university professor were futile, given his "racial" background. In 1938, threatened by the Nazi political regime, he left Germany under the pretext of taking a vacation. He went to England, where he was soon followed by his wife and young son, who was then only two and a half years old. The family was not to return to Germany. Sadly, Walter Bodmer told us that his father never recovered psychologically from the ostracism he faced in Germany, Nazi or not.

Much later, and like Cavalli, Walter Bodmer became fascinated by both mathematics and genetics. He obtained degrees in both from Cambridge University and worked as a graduate student with R.A. Fisher. It is in Cambridge, in 1957, that Cavalli first met Bodmer in Fisher's laboratory.

Collaborating at Stanford with Walter Bodmer

In 1961, Bodmer was hired as a postdoctoral fellow at Stanford University by Joshua Lederberg . As explained earlier, bacterial sex in *E. coli* had already been discovered by Lederberg and largely unraveled by Cavalli and the Lederbergs, so there was little motivation to continue in this line of work. Under the impulse of Lederberg, Bodmer became interested in another way that bacteria have developed to exchange genes—**transformation**.

Sexual reproduction, including that in bacteria, consists in sharing genes. We saw at length in chapter 2 how *E. coli* does so. However, some bacteria, including *Bacillus subtilis*, Bodmer's pet organism, exchange genes differently. In this species, DNA added to their culture medium can be spontaneously picked up by *B. subtilis* cells. This is a very rare natural phenomenon, and while *Bacillus* species have developed this DNA uptake ability, the enormous majority of other bacterial genera have not. Thus, *B. subtilis* cells can acquire new gene variants simply by incorporating DNA provided to them. Why did Bodmer concentrate on transformation right after his arrival at Stanford? It turns out that Joshua Lederberg and others had hoped that a transformation system could be developed for *E. coli*, a model organism much better understood than

B. subtilis. Unfortunately, early efforts at transforming *E. coli* were unsuccessful, so Lederberg asked Bodmer to turn his efforts to *B. subtilis* instead. It is not until 1972, well after Bodmer left Stanford, that *E. coli* proved after all to be transformable by purified DNA.

So, interestingly, Bodmer did bacterial genetics that involved gene exchange among cells, and had training in mathematics. Cavalli, of course, was initially a bacterial geneticist who had become proficient in mathematics. It is easy to see the synergy between these two men. And synergy there was. Cavalli was invited to Stanford in the summer of 1962 to team-teach a genetics course with Bodmer, who, in the meantime, had been offered a professorship. Before that, Cavalli had already been introduced to upscale Palo Alto, powerful Stanford University, and the great California weather in 1954 and then again in 1960, to teach a course on human population genetics upon the invitation from Joshua Lederberg.

Their common interest in mathematics and genetics led Cavalli and Bodmer to write a massive textbook on human genetics, *The Genetics of Human Populations*, which came out in 1971. In 1976 they also published the equally massive *Genetics, Evolution, and Man*, and several articles analyzing in great detail mathematical models for migration and genetic drift. To work on the books, Cavalli spent a sabbatical year at Stanford in 1968–69 (fig. 3.8), and Bodmer spent several periods of time in Italy. Bodmer left Stanford Uni-

Figure 3.8 Cavalli, with Sir Walter Bodmer (*left*) (Stanford, California, 1968–69). (From the collection of L. L. Cavalli-Sforza)

versity in 1970, just one year before Cavalli returned to Stanford to take a permanent position in the same department in Bodmer's former laboratory. As Cavalli put it to us, "I learned math from Bodmer and he learned human genetics from me."

Bodmer went back to England, became a professor at Oxford University, then went to London to direct a large cancer research laboratory, returned to Oxford again, and was the recipient of many honors and honorary degrees. From 1990 to 1992 he was also president of the Human Genome Organization (HUGO), an organization of scientists started effectively as a society for those with an interest in the human genome. Numerous scientists, many of them members of HUGO, ended up sequencing the 3.1 billion base pairs of the human genome in the year 2000. At the same time that he was president of HUGO, in 1991 Bodmer asked Cavalli to start the Human Genome Diversity Project (which we discuss in chapter 7). Therefore, Bodmer and Cavalli have interacted for many decades, starting at Cambridge University in the 1950s, then culminating at Stanford University in the 1960s (including a trip to Africa together to study the Pygmies), and continuing to date.

But, as we explain next, the normally placid world of academia known by Cavalli until the time he considered going to Stanford was to undergo some drastic changes in the late 1960s. In the next section we analyze Cavalli's move to Stanford, curiously bracketed by a social revolution in Europe that had its roots in university ivory towers and well-intentioned irate students.

The 1968 Student Revolution in Europe

In 1958, Joshua Lederberg, Cavalli's former collaborator, received the Nobel Prize for Medicine or Physiology for his contributions to bacterial genetics. The next year he accepted a professorship at Stanford University and was to become instrumental in Cavalli's eventual permanent move there. Shortly after he arrived on campus, Lederberg offered Cavalli a position at Stanford to help organize the genetics department. Cavalli did not accept the position at that time, but the offer was left open for many years. As mentioned earlier, Cavalli was invited to teach a course at Stanford in summer 1960 and in 1962 he was asked to repeat the course (which he then cotaught with Bodmer). With a Stanford option now more firmly in mind, Cavalli next spent a "trial" year there in 1968–69 with his family, now consisting of his wife Alba, sons Tommaso (a future M.D. and World Health Organization officer) and Francesco (a future writer and movie director), and daughter Violetta (a future professor of computer science). Son Matteo, a particle physicist who was working at the time at the Stanford Linear Accelerator (SLAC), was already in Palo Alto. This trial year meant that Cavalli escaped much—but not all—of the aftermath of the 1968 student revolution that rocked Italy, France, and Belgium.

It all started in France, in the beautiful month of May. Gen. Charles de Gaulle had been in power for years, and his conservative and nationalistic policies had started tiring a significant segment of the French public, including many university students, workers, and the trade unions. Of course, the more vocal factions were the ones situated on the left and extreme left, the socialists and the communists. French higher education had become fossilized, and professors were called "mandarins." In other words, many students now saw their teachers as being not much more than potentates equipped with academic credentials.

And then it all exploded. Students invaded and occupied the Sorbonne in Paris, and the Nanterre campus, a branch of the University of Paris specializing in teaching and research in social studies. The students had come up with a beautiful slogan: "L'imagination au pouvoir" (Power to the imagination), indicating by that that they wanted deep changes in the institutions of government and higher education. They were not simply protesting exams or homework, they wanted to change society. They also built barricades in the streets of Paris. This was the 1871 Paris Commune all over again. President de Gaulle's police forces reacted with extreme brutality, and the students responded in kind. Police cars were set on fire, and the city became saturated with tear gas. Students were arrested and sent to jail in droves, but not without putting up a fight. A few days later, the workers went on strike and France became completely paralyzed. The revolution was making progress.

Meanwhile, news of the new French Revolution had spread like wildfire to two of its neighbors, Italy and the French-speaking part of Belgium. In Italy, the discontent with sclerotic academic institutions was aggravated by the conservative policies of the ruling party, the Democrazia cristiana. University campuses were occupied, barricades went up, and police charged. Students spent entire nights in "free assemblies" redefining the nature of higher education and, in particular, what it meant to be young, privileged, and socially aware—all this in the midst of a dissatisfied proletariat. Like their counterparts in France, Belgian students chanted, "Ce n'est qu'un début, continuons le combat" (This is only the beginning, let's continue the struggle). Undoubtedly, their Italian brothers and sisters in the revolution said something similar. Basically, they were all Trotskyists, anarchists, or anarcho-syndicalists, as indicated by their flags—red, black, and of course red *and* black. In many ways, the revolutionary students of France, Italy, and Belgium were emulating the forces (which included the Lincoln Brigade, from America) and the political factions which had opposed the attack of the fascist Gen. Francisco Franco against the Spanish Republic in the late 1930s.

Professors in the three countries were of course not amused. Their authority had been challenged and was now nonexistent because campus buildings were occupied by student revolutionaries (most of the time without serious damage to university or personal property). Did the students change the world? No. In spite of their enthusiasm, rejection of social classes, and resilience, revolutionary students could not convince the proletariat that the old ways of doing things should be discarded. Gradually, worker support disappeared and was replaced by hostility. Traditional political parties, even the communists, rejected them. The forces of conservatism and reactionaryism won the day, and the revolution petered out only several months after it had begun. Eventually, the dream of a new social justice died and students went back to their studies, but not without having softened the "mandarin" status of their professors, a limited and rather bitter victory. Curiously, a similar student movement also took place in the United States at roughly the same time. There, however, it was not basically motivated by social injustice and class consciousness; the Vietnam War was at its core. Nevertheless, some parallels between the two uprisings can be drawn.

Cavalli brought up the student revolution in a conversation with us over lunch because his son Matteo had been a leader of the student revolution at the University of Pavia, and his second son, Francesco, had participated in it as a student at Berkeley. Francesco, easily identified by his long hair in the late 1960s, was even beaten up by the police and, later, by rednecks who nearly killed him on a hitchhiking trip from San Francisco to New York. Nonetheless, it seemed that Cavalli had little sympathy for the student movement. He said that he was in fact very favorable to the movement in Italy in the beginning (1968) when it was focused on university changes and reform. He lost interest when the movement became politicized and the students tried to "change the world," an infinitely more ambitious goal whose achievement Cavalli doubted.

What happened at the Laboratorio Internazionale di Genetica e Biofisica (the LIGB, now called IIGB, where the "I" stands for Istituto) in Naples is a good example of what Cavalli found irritating with Italy, both before and after the student revolution. The LIGB had been founded in 1963 by Buzzati-Traverso, Cavalli's mentor. It quickly became a renowned institute for biological research, attracting many famous British and American scientists. But then, in 1964, the LIGB and other research institutions were caught in the middle of a power struggle between the Christian Democracy (the DC) and the Socialist Party, which resulted in the politically motivated jailing of the directors of two major Italian research organizations. Buzzati-Traverso, director of the LIGB, offered his resignation in protest. Cavalli was then asked to become interim director of the LIGB, his firm intention being to reinstate Buzzati-Traverso as permanent director as soon as possible. After seven months of diplomacy, he succeeded, and the LIGB was saved from potential disappearance. One can easily imagine that the LIGB had become close to Cavalli's heart.

Unfortunately, four years later, in 1969, the LIGB nearly self-destructed. This time, Maoist researchers and technicians, inspired by the 1968 student revolution, stopped working and spent most of their time calling up "general assemblies" to discuss the politics and ethics of scientific research. In typical Maoist fashion, they decided that all research projects should be aimed at the welfare of the workers. Not only that, they also instituted a Maoist-like program of "reeducation," in which researchers and technicians had to spend equal time washing test tubes and doing research. It is at this point that Buzzati-Traverso resigned in despair, and for good. Fortunately, after much bickering and politicking, the IIGB today is back to normal.

In the end, said Cavalli, the 1968–69 student movement had just been a gigantic waste. Francesco respectfully disagreed with his father's opinion. All in all, it may be that Cavalli's detachment from things political is due to the fact that, as cited at the beginning of this chapter, he is "nearly glued" to science.

In any case, Cavalli's experience at Stanford during his trial year and later allowed him to form impressions of science in the United States as compared to elsewhere. Earlier we noted his view that science in England is highly efficient. This, he said, is because researchers there are allowed their own niches, resulting in little redundancy. In Cavalli's view, the United States, although good for doing research in terms of facilities, money, and the availability of collaborators in many fields, suffers from the "rat race," the drive to be first. This not only creates unnecessary anxiety but also results in many scientists doing the same thing, hence a wasted duplication of effort.

We now start a new chapter, one that some claim was the most significant in Cavalli's life. The year is 1971, Cavalli is forty-nine years old, and he is moving his family and belongings across the Atlantic to the North American continent, where he will settle down in Palo Alto. For good. And here, right at this point of transition between his life in Italy and in the United Staes, Cavalli begins some new and very interesting excursions into human culture.

Chapter 4

Excursions into Human Culture (1970–)

The move from Italy to Stanford in 1971 was an important decision in Cavalli's career. It was a move he has not regretted, although he did report missing the quality of Italian food and the beauty of small Italian cities. In his view the decision to leave Italy to live in the United States was probably more difficult for his wife, Alba. Most of his children were grown by this time; only the youngest, Violetta, then age twelve, was raised in the United States. Cavalli had claimed that the family's 1968–69 stay at Stanford was a "trial year" on which to assess whether to move permanently, but in fact he did not tell Alba that this was a trial or that he was thinking of making a permanent shift. This, he said, he kept from her in order to make "the experiment" a true test and not bias the results, much in the same way that you do not tell subjects whether they are taking the placebo or the real drug in medical experiments. In any event, he was happy that Alba agreed to the move.

At Stanford, in addition to his fast-developing research, Cavalli also taught a few courses between 1971 and 1992. In 1992 he became professor emeritus at age seventy. For over a decade he taught a course in "Human Genetics" and gave some lectures in a course in "Medical Genetics." Even after becoming emeritus, he taught a course called "Genes, Peoples, and Languages" for a few years. He describes his own teaching as "not bad but not great." He thinks, however, or at least hopes, that like good wine, his lecturing has improved with age. Another close observer told us that his lecturing style was not always effective since Cavalli "has too much to say and the digressions at the beginning of the talk become the talk itself."

Both before and especially after his move to Stanford, Cavalli's career developed its unique combination of research on population genetics with excursions into human culture. There are three major sources for his interest in culture. One, discussed in the previous chapter, was his field experience with the hunter-gatherer populations of Africa and, through them, his fascination with human cultural differences and similarities. A second source was his growing belief that human cultural distributions (i.e., human languages, human cultural traits, and remains of artifacts) could be used in conjunction with each other and with genetic data to confirm, reject, or modify hypotheses about human prehistory. Yet another source of his interest in culture was the idea that the concept of human cultural learning was a valid weapon against racist arguments that differences between people (for example, different IQ scores among ethnic groups) were due to biologically determined "racial" differences. This chapter shows how, from these sources, Cavalli's excursions into human culture led to some of his specific contributions to anthropology.

The Neolithic in Europe

One day in 1963, Cavalli happened to visit the Pigorini Prehistoric and Ethnographic Museum in Rome. There he saw reconstructions of large stone (megalithic) buildings from Puglia in Southern Italy. It struck him that these structures resembled the unusual megalithic buildings (called *nuraghi*) he had earlier seen on Sardinia, the large island off the Italian coast. He wondered: Could these buildings have been constructed by the same people? And could genetic comparisons between contemporary Puglians and Sardinians show a link? This was the first gene/archaeology "light bulb" to turn on in his mind. Little did he know at the time that this thought would lead him to a new theory of the spread of agriculture in Europe, to collaboration with American archaeologist Albert Ammerman, to a controversy (which turned out to be a pseudo controversy) with geneticist Bryan Sykes, and, ultimately, to a grand reconstruction of world prehistory using archaeological, linguistic, and genetic data (see chapter 5).

The first thing Cavalli did with his new ideas was to test for genetic connections between Puglians and Sardinians. He collected blood samples from Puglia and compared these to genetic information on Sardinians. The result was negative: there were no significant genetic similarities between these two groups to support the idea that the same people built the megalithic structures in both areas. Undeterred, Cavalli wondered about other situations where matching genetic and archaeological data could throw light on human prehistory. He decided that one such case might be the development of agriculture in Europe. He realized he needed an archaeologist with whom to collaborate; in time he found Albert Ammerman, who in 1970 happened to be in Italy. Cavalli was in Pavia at the time; he invited Ammerman there and the two began their collaboration, both at Pavia and later at Stanford University.

This collaboration focused on a major transition in European prehistory. Archaeologists divide human prehistory into periods based on stone tool (lithic) technology. There are three major periods: Paleolithic, **Mesolithic**, and

Neolithic. The Paleolithic (old stone age) is a vast period from the beginning of stone tools, about 2 million years ago, to the end of the last Ice Age about 15,000 years ago. In Europe the Paleolithic covers the period of *Homo erectus* (known as the Lower Paleolithic), Neanderthals (Middle Paleolithic), and *Homo sapiens* (Upper Paleolithic), represented by the ancestors of the Cro-Magnon peoples as early as 40,000 years ago. The Mesolithic (middle stone age) in Europe spans the period from the last Ice Age to the appearance of agriculture, or the Neolithic, between 11,000 and 5,000 years ago, depending on the location. The Mesolithic is associated with a certain advanced stone tool technology; in the Neolithic, or agricultural, period a completely new technology appeared, which included sickles, grinding stones, and pottery, all associated with agriculture.

As has been known for many years, agriculture was independently developed in several places in the world, the first being the Middle Eastern Fertile Crescent (covering sections of modern-day Syria, Turkey, Iraq, and Iran) and spreading to Europe from there. A study published in *Science* in 2000 suggests that the very first beginnings of agriculture can be pinpointed to a small area (the Karacadag region) within the Fertile Crescent—a core area in southeast Turkey and northern Syria in the upper region of the Tigris and Euphrates Rivers. Evidence for this region as the cradle of agriculture is that the wild varieties of all seven of the original Middle Eastern domesticated crops (i.e., einkorn wheat, emmer wheat, barley, lentil, pea, bitter vetch, and chickpea) are found only in this core area. In human prehistory the Neolithic is an extraordinarily important transition, bringing not only a dramatic change in subsistence but, with that, a sedentary way of life, population growth, new diseases, social inequality (including a general lowering of the status of women), and ultimately the development of cities and the rise of the state.

In the 1970s Cavalli and Ammerman developed the theory that agriculture spread to Europe from the Middle East mostly in a process they termed "demic diffusion"—that is, it spread through migrations of farmers themselves. This theory was opposed to the then current notion in archaeology that farming in Europe spread through **cultural diffusion**—the knowledge of farming (but not the farmers themselves), spread out from a point of origin. In cultural diffusion, ideas and practices spread from group to group without major population movement, and they spread because they are seen as advantageous to those exposed to them. This notion that demic diffusion also may have played a role in the spread of agriculture in Europe was at this time a radical new idea. Although Ammerman and Cavalli thought it likely that both processes (cultural diffusion *and* demic diffusion) took place, they did not know their relative importance.

Why did Cavalli and Ammerman think that the spread of agriculture in Europe was demic? Cavalli explained his thinking in terms of his observations of African hunter-gatherers (covered in the last chapter). "I observed demic diffusion firsthand," he said, with reference to the hunter-gatherers and their agricultural Bantu neighbors. What Cavalli had seen is that the hunter-gatherers did not take up agriculture after being exposed to it by the Bantu. They are doing so now, as their remaining forest is rapidly being destroyed. But before that, they did not see agriculture as a superior or advantageous way to secure food. They were quite happy with their hunting-gathering mode of life, to which they were well adapted; they saw that agriculture required more work and left less leisure time. What is happening, however, is that Bantu agriculturists are themselves spreading through the tropical forest of the Pygmies, taking their agriculture with them. Their population is also expanding at a higher rate than that of the hunter-gatherers. In time the hunting-gathering way of life will be gone and agriculture will have spread through the tropical forest. This is demic diffusion of agriculture. Cavalli and Ammerman reasoned that something similar might have occurred in European prehistory.

Ammerman and Cavalli's first step to test their demic diffusion hypothesis was to mark on a map of Europe the dates for the first appearance of agriculture at various sites, using carbon 14 dates of domesticated cereals. They saw that agriculture took about 4,000 years to spread through Europe, beginning about 9,000 years ago from a point of origin in the Middle East (fig. 4.1). They

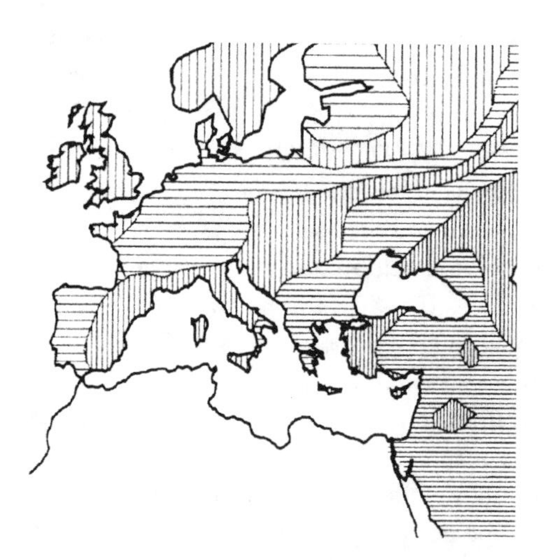

Figure 4.1 The spread of agriculture in Europe based on radiocarbon dating of Neolithic sites. Oldest sites (8,500 years before present or older) are found in the Middle East and Anatolia. The most recent (less than 6,000 years before present) are located in the British Isles, Iceland, Scandinavia, Western France, Northern Germany, and Russia. The map clearly shows a gradient (in time and space) between the oldest and the newest sites. (From Cavalli-Sforza and Minch 1997:249. Reproduced with permission.)

calculated the average rate of **expansion** at one kilometer per year. They saw, then, a slow but steady spread.

Next, Ammerman and Cavalli applied R. A. Fisher's (1937) "**wave of advance**" model to their data.[1] Cavalli, as we saw in chapter 2, worked with Fisher at Cambridge in the 1940s and was among the first to apply Fisher's work to demographic contexts. This "wave of advance" was originally a model for the spread of an advantageous gene through a hypothetical population, given certain assumptions. Cavalli and Ammerman felt that the same wave model could be applied to the spread of farming in the Ammerman-Cavalli hypothesis of demic diffusion. (It can also be applied to many other situations—for example, the spread of a rumor or an epidemic.) They also made use of others' mathematical demonstration that if a population increases with a slow migration outward in all directions, it will move in a wave of expansion that progresses at a constant rate. When they put their data into a wave of advance model, they found that the demography was compatible with the theory of demic diffusion.

A computer simulation of the wave of advance theory applied to spreading Neolithic farming populations in Europe is shown in figure 4.2. In this

---

1. For mathematically inclined readers, Fisher's wave of advance equation applied to an expanding population is written

$$dp/dt = ap(1 - p/M) + m(d^2p/dx^2)$$

where the signs d and d^2 represent first and second derivatives, p in this case is the local population density, t is time, m is a migration coefficient representing the distance migrated per individual per unit time, a is the growth rate of the population, M is the maximum population density at saturation, and x is a space coordinate. The rate of advance (ρ)—that is, the rate at which the wave front moves in time and space—is represented by the equation

$$\rho = \sigma\sqrt{2\alpha}$$

where σ is the standard deviation (the spread) of the migration, α is the growth rate of the advancing population, and $\sqrt{}$ is the square root.

The wave of advance theory can be used to understand how various, advancing early populations of *H. sapiens* acquired different gene combinations that can still be traced today. In 2003, Cavalli and coworkers Chris Edmonds and Anita Shen Lillie demonstrated by using computer simulations that gene mutations, even in the absence of natural selection, when they occur on the wave front of an expansion, may travel with the wave front and reach high frequencies far from their place of origin, because of the combination of a strong bottleneck and population growth. These simulations agree very well with the observation that geographically distinct populations (presumably originating from an ancestral population, a portion of which stayed behind and portions of which migrated in different directions) display distinctive gene combinations, especially in regions farthest from the origin of expansion (also see chapter 6).

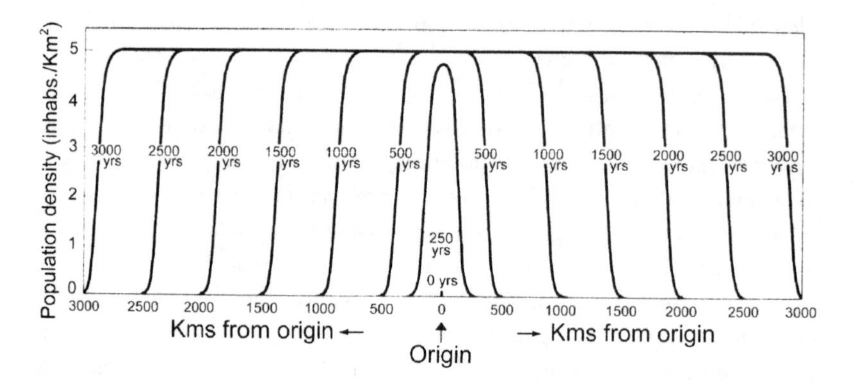

Figure 4.2 The wave of advance theory applied to a spreading population that reaches demographic saturation at 5 inhabitants per square kilometer, a growth rate of 3.9%, and an original population size of 0.1 inhabitant per square kilometer. The wave front is represented 250, 500, 1,000, 1,500, 2,000, 2,500, and 3,000 years after the start of the migration. Y axis = population density; X axis = distance reached in kilometers. (Redrawn from Ammerman and Cavalli-Sforza 1973:352)

simulation, the population starts at a density of 0.1 inhabitant/km^2 (origin) with an assumed population growth rate of 3.9 percent per year and saturation of the land reached at a population density of 5 inhabitants/km^2. It can be seen that, under these conditions, farmers started expanding from their origin in northern Syria around 9,000 years ago and took about 4,000 years to cover 4,000 km, at a rate of 1 km/year. This number is consistent with radiocarbon dating data.

It is reasonable to assume that these Neolithic people moved at a demographic rate comparable to that of some people in the developing world today. Other researchers have recently calculated that the migration rate observed over a two-year period in an Ethiopian group living with primitive farming techniques fitted exactly the observed rate of expansion of agriculture in the Neolithic.

Although suggestive, all this still did not prove that the spread of farming was most likely due to demic diffusion as well as to cultural diffusion. It was here that Ammerman and Cavalli sought genetic data to better test their hypothesis. At first they used just a few genes—for example, the *Rh–* and *Rh+* genes. Distributions of frequencies for these genes confirmed their idea of demic diffusion, but they realized that data from one or even a few genes was not enough to substantiate their hypothesis; in fact, the more genes they could use, the better the test would be. Later they added data on more genes, at one point using 39 and later increasing this number to 95. Their data bank included information on human leukocyte antigen (HLA) genes which had been extensively studied by Walter Bodmer and his wife Julia. A breakthrough came in 1978 when Cavalli and his Italian colleagues, Alberto Piazza and Paolo Menozzi, fed a great deal of genetic data into a computer (see fig. 4.3). They performed a principal component analysis (see chapter 3) of the data and achieved a stunning result: the genetic map of Europe showed a perfect match with the map of the spread of agriculture! Figure 4.4 gives the first (and most significant) principal component clearly showing a genetic gradient in Europe.

In other words, the genetic gradient of Europe turned out to be exactly what one would expect if one supposed that agriculture was spread by migrating farmers from the Middle East into Europe over a 4,000-year time period, with a gradual expansion at a rate of about 1 kilometer per year. Ammerman and Cavalli's demic diffusion model had also assumed that the migrating Neolithic population was expanding at a faster rate than that of Europe's Mesolithic populations and that there was some genetic admixture of the two populations. Eventually the genetic results of Cavalli, Piazza, and Menozzi were confirmed by population geneticist Robert Sokal.

Figure 4.3 *From left to right*: Cavalli, Paolo Menozzi, and Alberto Piazza (Italy, 1978). (From the collection of L. L. Cavalli-Sforza)

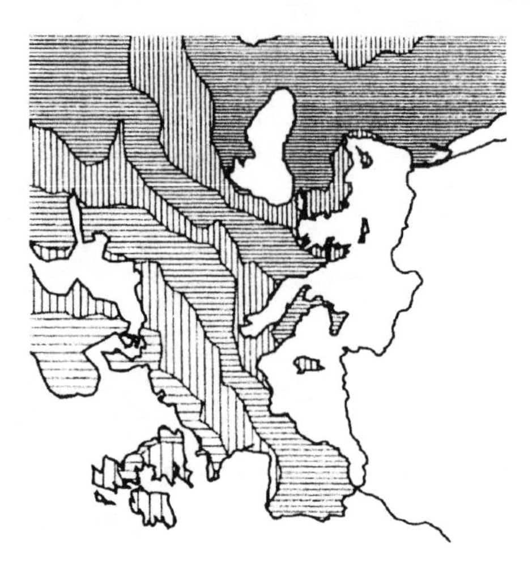

Figure 4.4 The first principal component of the frequencies of 91 genes in Europe. A gradient between the Middle East and Northern Europe is clearly visible. The scale is arbitrary. Compare this genetic gradient (in space) with the radiocarbon gradient shown in figure 4.1. (From Cavalli-Sforza and Minch 1997:249. Reproduced with permission.)

From Ammerman and Cavalli's work, a clear picture of the Neolithic in Europe was emerging. Neolithic farmers in the Middle East were expanding in population as agriculture brought greater and more regular supplies of food. This expansion meant that they themselves needed to move outward, albeit gradually, to avoid exhaustion of resources in one region and population saturation. As they expanded into Europe, they came into contact with Mesolithic groups. Cavalli and Ammerman supposed (and archaeological data confirms) that these two groups were not in conflict. Rather, they occupied separate ecological niches. The Mesolithic hunter-gatherers stayed in forests to pursue game; the farmers went to woodlands which they cleared to grow crops. Yet the two groups were in contact and, as is inevitable under contact, there was some genetic admixture. But the Neolithic populations were expanding at a faster rate. Ammerman and Cavalli surmised that in Europe agriculture gave the Neolithics an edge because their food could be stored for use during winter, a time when the hunter-gatherers faced food shortages that acted as checks on their population growth. With time, the Neolithics and their farming practices had spread everywhere in Europe, and the hunting-gathering mode of subsistence was virtually at an end.

Most European archaeologists came to accept the Ammerman-Cavalli hypothesis of the demic diffusion of agriculture in Europe, but acceptance took time and effort. Anglo-American archaeologists took more time, and even today some are still uncertain. They had been trained under the idea that nobody ever moved, except traders, an ideology Ammerman baptized as "indigenism." As evidence of the general resistance to the idea of the demic diffusion of agriculture in Europe, Ammerman and Cavalli had trouble publishing their first paper on the topic; one peer reviewer for a journal sent back the discouraging comment, "This is impossible." Indeed, Cavalli told us that his work on the demic diffusion of agriculture has been one of the most debated of his scientific contributions. Many people, he said, simply did not understand that one would not see the genetic gradients (resulting from the farmers' expansion) we do see in Europe today if Middle Eastern Neolithic farmers had not gradually migrated into Europe and mixed with local Paleolithics. Many years after Cavalli's and Ammerman's original proposal, the acceptance of demic diffusion was boosted by the support of archaeologist Colin Renfrew. Renfrew showed how demic diffusion could account for the spread of Indo-European languages (which he claimed originated in Anatolia, Turkey) to and through Europe, where they are spoken today.

But soon, another debate was to be spurred by Cavalli and Ammerman's work, this one having to do with the magnitude of the expansion of farmers into Europe. This controversy was ignited by Bryan Sykes, a professor of genetics at the Institute of Molecular Medicine at Oxford University. Sykes himself is well known for a number of achievements. For example, he extracted DNA from ancient human bones, he participated in the genetic analysis of the last of the royal Russian Romanovs (helping to disprove the famous claim of Anna Anderson that she was Anastasia), and he examined the Italian Ice Man's remains. He has also written an entertaining book, *The Seven Daughters of Eve* (2001), in which he disputes Cavalli's work on demic diffusion.

Sykes has also contributed to the unraveling of human prehistory through his work with mitochondrial DNA (see chapter 3), which, as we recall, passes to children exclusively through the mother over the generations. Examining several hundred DNA samples from all over Europe, Sykes and his research team claimed to have found seven distinct clusters. They further claimed that dating of the sequences showed that all but one cluster were older than 10,000 years, meaning that they were genetic echoes from ancestors who lived in Europe *before* the spread of agriculture. For Sykes this meant that the so-called demic diffusion of Neolithic farmers into Europe from the Middle East could not have been a major migration. Initially his group's data showed indeed that demic diffusion would at most account for only 4 percent of the European genes. Later on, their estimate was revised and increased to 20 percent after Cavalli and Eric Minch showed that the results of Sykes's team were not statistically significant.

Meanwhile, Sykes claimed that his own reformulation of European prehistory goes against the now-current dogma that Europe's hunter-gatherers were overwhelmingly replaced through the migration of the Neolithic farmers. Here is where Cavalli's work comes in, or so Sykes thought it should come in. In fact, however, Sykes misleadingly tied Cavalli to this notion that the expansion of farmers into Europe was "overwhelming." Sir Walter Bodmer told us that Sykes was misled on this whole issue because he never understood the idea of the "wave of advance."

In reality, all of Cavalli's publications on the topic (up to the time of Sykes's own mitochondrial DNA publications) were addressing the question: *How* did farming spread to Europe? Cavalli was not, at least not back then, asking: *How major* was this expansion? Indeed, until he directly commented on Sykes's own work in a 1997 publication with Eric Minch, Cavalli and his coauthors never said what percentage of modern Europeans share genes with the Neolithic farmers. Yet in his book Sykes ties the "overwhelming expansion" idea directly to Cavalli:

> It was Luca who first formulated the theory which had come to dominate European prehistory over the preceding quarter century. According to this theory, or at least the version believed by archaeologists, farmers from the Near East had overwhelmed the descendants of the Cro-Magnons, who themselves had replaced the Neanderthals. This was a large-scale replacement which meant that most Europeans traced their ancestry back not to hunter-gatherers but to farmers. (*The Seven Daughters of Eve*, 146)

Sykes then claimed that his mitochondrial DNA studies revolutionize European prehistory. He announced to the world that the Neolithic expansion into Europe could not have been "overwhelming" and, with a bit of dramatic flair, he wrote that his genetic findings represent "not the faint whispers of a defeated and sidelined people but a resonant and loud declaration from our hunter-gatherer ancestors: 'We are still here'" (2001:150).

To be sure, Cavalli and his coworkers are not solely to blame for the "overwhelming migration" idea, according to Sykes. Misguided archaeologists were also at fault. Part of the problem, wrote Sykes, was that Cavalli and his team had described the Neolithic expansion in terms of R. A. Fisher's mathematical model of the "wave of advance." In time the term "demic diffusion" was replaced with the classier term "wave of advance," so that "it may just have been that archaeologists were beguiled by the power of the phrase 'wave of advance'" (2001:148). Sykes continued: "The idea of a gradual influence of incoming agriculturists had been replaced in the collective psyche by the image of an unstoppable tidal wave of land-grabbing farmers that swept away everyone and everything in its path" (148).

Sykes then wrote about a direct confrontation between himself and Cavalli, along with Walter Bodmer, Cavalli's coauthor of the 1976 book *Genetics, Evolution, and Man*. This was in 1995 at a Euro-conference on "Population History" in Barcelona, Spain, where Sykes was giving a paper on the about-to-be published results of his team's mitochondrial DNA studies and their implications for European prehistory. Sykes reported that the audience was large, with Cavalli and Bodmer sitting side by side in the front row. Then,

> As I went from one point to another I could see that Walter was getting agitated. He began to mutter to himself, then to Luca; at first inaudibly, then louder and louder. 'Rubbish,' 'Nonsense,' I thought I heard him say. . . . As I came to the concluding slide, I could see the steam coming out of his ears. (*The Seven Daughters of Eve*, 151)

Sykes claimed that at the end of the talk Bodmer denied that he and Cavalli had ever said that the farmer migration into Europe was an overwhelming replacement of the hunter-gatherers. But Sykes was evidently prepared. He claimed that he whipped out a copy of *Genetics, Evolution, and Man* and read out the very page where this allegation is made!

Sykes could hardly contain his excitement: "There it was in black and white. This was massive replacement in all but name" (2001:152).

Now there are a number of interesting aspects to this story and its aftermath. First, we were very intrigued to hear from Cavalli that he has no recollection whatever of Sykes reading out the page from his and Bodmer's book! Sir Walter Bodmer told us that he, too, has no recollection of Sykes reading from this book during his talk. Second, here is the passage from Bodmer and Cavalli's 1976 book that Sykes claimed to have read out loud:

> "If the population of Europe is largely composed of farmers who gradually immigrated from the Near East, the genes of the original Near Easterners were probably diluted out progressively with local genes as the farmers advanced westward. However, the density of hunter-gatherers was probably small and the dilution would thus be relatively modest." (Quoted in Sykes, *The Seven Daughters of Eve*, 151–52)

Now here is the same passage from Bodmer and Cavalli's book, originally embedded in other sentences, showing what Sykes chose to leave out:

> "The increase in population size of farmers and the mobility associated with the techniques of shifting agriculture may well have caused a diffusion of people ('demic' diffusion) rather than the idea ('stimulus' or

> 'cultural' diffusion). *Probably both phenomena have occurred, but their relative importance remains a problem.* . . . If the population of Europe is largely composed of farmers who gradually immigrated from the Near East, the genes of the original Near Easterners were probably diluted out progressively with local genes as the farmers advanced westward. However, the density of hunter-gatherers was probably small and the dilution would thus be relatively modest. . . . *The problem of the relative importance of demic and cultural diffusion of agriculture, however, is not yet settled and requires further investigation.*" (*Genetics, Evolution, and Man*, 548; emphasis added)

There it is, in black and white.

Third, in a 1997 issue of the *American Journal of Human Genetics*, Cavalli and Minch wrote a critique of the work of Sykes's team which had been published in the same journal in 1996. Here, for the first time, Cavalli and Minch offer their own estimate, based on Cavalli's genetic data, of the percentage of modern Europeans who share Neolithic farmer genes: 26 percent—not far from Sykes's own estimate of 20 percent, based on mitochondrial DNA. With this, there really was not much of a controversy after all. Sykes himself noted the agreement and was pleased with the result, but by this point in Sykes's book, Bodmer and Cavalli have already served his purpose. Sykes had portrayed Bodmer and Cavalli as members of the cranky old establishment, refusing to see the error of their ways or even admitting their role in perpetuating the "overwhelming migration" hypothesis, merely two of many other misguided scientists and other obstacles Sykes needed to overcome in his self-led journey to find the real truth of European prehistory.

The saga of the "overwhelming migration" idea, however, does not end here. A more recent study using Y chromosome data gives a far higher percentage of Neolithic genes in the European population (as we cover in chapter 6).

Most of Cavalli's work on the demic diffusion hypothesis took place after his move to Stanford in 1971. At Stanford, Cavalli was hired as a tenured full professor, and he served as chair of the genetics department for four years, from 1985 until 1989. As he told us, he had requested at the time he was hired that he not be asked to be Department Chair. Nevertheless, he agreed to serve, out of a sense of duty, during a difficult time for his department.

In this same period Cavalli also embarked on another contribution to anthropology, a theory of cultural evolution. This involved him in an important, long-term collaboration with Stanford biologist Marcus Feldman, resulting in a number of joint publications. With this collaboration, Cavalli significantly furthered his thinking about culture and cultural evolution and furthered his interests in mathematical representations of cultural processes.

A Theory of Cultural Evolution

A year after arriving at Stanford, Cavalli gave a seminar to a group of scholars interested in mathematical biology. Here he addressed some of his ideas about a new topic on his mind—analogies between biological and cultural evolution. Biological evolution works through the effects of natural selection on mutations, and in Cavalli's view something analogous might be occurring at the cultural level—i.e., *cultural mutations* (such as innovations) arise and undergo a process of *cultural selection* (acceptance or rejection) through choices made by individuals. In addition to these analogies, Cavalli had also written about *cultural drift* and noted the similar role that migration could play in both biological and cultural evolution. He had further brought up mechanisms of cultural transmission, noting that some of these, compared to biological inheritance, could result in rapid or radical change. After covering cultural evolution in his seminar, Cavalli presented some mathematical modeling of these ideas.

Sitting in the audience listening to Cavalli's talk was Marcus Feldman, a mathematician and population biologist from Australia, who had recently joined Stanford University as an assistant professor. Feldman had been a graduate student at Stanford (with Bodmer as a member of his Ph.D. committee), where he had met Cavalli during Cavalli's earlier visits to that campus. At the end of the talk, Feldman asked Cavalli a few mathematical questions. Cavalli was very interested in these and suggested the two of them discuss the problems further. When Cavalli's presentation was over and the room cleared, "We sat down right there and solved a few of the mathematical problems," Feldman told us. "And that's how it all began." In addition to their friendship, this encounter led to a long and close collaboration between Cavalli and Feldman and resulted, among other joint publications, in their book *Cultural Transmission and Evolution: A Quantitative Approach* (1981). Over the next year, Feldman and Cavalli met three or more nights a week at Cavalli's house, carrying out mathematical computations and fleshing out ideas about cultural transmission and change.

Before all this was going on, another development had been taking place at Stanford that further inspired Cavalli and Feldman's work. Arthur Jensen, professor of education at the University of California, Berkeley, had published an article stating that the difference in intelligence (measured in IQ testing) between black and white people in the United States was hereditary and therefore unchangeable. William Shockley, a physicist at Stanford and winner of a joint Nobel Prize for inventing the transistor, supported Jensen's ideas and organized a number of conferences to propagate them. Indeed Shockley went a step further: he proposed that black women be given a monetary award of $5,000 in return for their voluntary sterilization!

This issue of race and IQ testing had started up in the late 1960s while Sir Walter Bodmer was still at Stanford and just before Cavalli arrived there. At that time (1967), a statement made by Shockley relating race and IQ differences was strongly criticized in a letter written by Joshua Lederberg and signed by Bodmer and others in the genetics department. Also, through one of Bodmer's postdoctoral students, Shockley was invited to come and talk with the Stanford geneticists about his ideas. Shockley came, bringing along Arthur Jensen. This event eventually led to the 1970 article written by Bodmer and Cavalli for *Scientific American*, countering Shockley's ideas.

Later, both Feldman and Cavalli became forceful participants in debates against Shockley at Stanford, pointing out that Jensen and Shockley completely ignored cultural learning (in this case, Jensen and Shockley had totally ignored, for example, that as late as the 1960s in the United States, blacks had very poor access to education compared to whites). Their participation in these debates with Shockley spurred them on to combine a theory of genetic evolution with that of cultural evolution. Although their book itself does not discuss the issue of race and IQ testing, it grew in part from Feldman's and Cavalli's desire to deflect arguments like those of Shockley's which failed to consider the influence of learning on human variation. The controversy surrounding the issue of IQ and race is fully discussed in Cavalli's book *The Great Human Diasporas*, and in his 1970 article with Walter Bodmer.

Feldman and Cavalli's close collaboration on evolutionary theory was for a time carried out in trying circumstances. In 1976, Cavalli developed bladder cancer. He was in the hospital for a few weeks and had major surgery, which then had to be redone. Cavalli often asked Feldman to come to the hospital so that the two of them could work on a number of their papers on evolution. Feldman told us they spent quite a bit of time together this way, "with Luca attached to tubes and things." This must have been not only challenging under the circumstances but grim as well. The doctors had given Cavalli only a 20 percent chance of surviving the cancer. Fortunately, he beat the odds.

Cavalli and Feldman's book made three contributions to culture theory. First, it developed ideas of cultural transmission and linked them to patterns of culture change. Second, it developed mathematical models of cultural transmission. Third, Cavalli and Feldman addressed the relationship between cultural and biological evolution. In all its aspects, the authors have emphasized the importance of culture in shaping human adaptations and variations.

As for what "culture" is, Cavalli and Feldman follow in their book a *Webster's* dictionary definition: "the total pattern of human behavior and its products embodied in thought, speech, action, and artifacts, and dependent on man's capacity for *learning* and *transmitting* knowledge" (1981:3; italics in original).[2] The emphasis is on culture as learned and transmitted knowledge and practice. So defined, culture is not unique to humans, but is most highly developed in this species. This definition of culture is useful to Cavalli and Feldman, but in order to build their theory—which involves analyzing the probability of specific cultural changes using an analogy with biological evolution—they do not focus on a "total pattern of behavior" but instead focus on discreet "cultural traits." Following Richard Dawkins, these are sometimes called "**memes,**" which can be any beliefs, values, or templates for behavior or the making of artifacts. On the other hand, Cavalli does not like the term *memes* because, in Dawkins's sense of the term, a meme is a culturally transmitted unit of *imitation.* This usage restricts cultural transmission to imitation and ignores other aspects of active transmission (such as teaching, for example). Cavalli has suggested other neologisms such as "mnemes" and "**semes**" but he now prefers the word *ideas*, which is more general. Cavalli notes that we do not know what the physical bases for ideas really are, other than the fact that neuronal circuits are involved. However, he also notes that much genetics was learned by scientists well before the physical basis for inheritance was understood.

At the core of Cavalli and Feldman's work is the idea that, just as the theory of Darwinian evolution was enhanced by Mendelian genetics, so too could the process of cultural evolution be enlightened by an understanding of how culture is transmitted. As they put it: "Accumulated experience with the study of biological evolution has taught us that central to a satisfactory theory of evolution is the sound knowledge of the laws of biological transmission. Similarly, knowledge of cultural transmission should be important in understanding cultural change" (1981:v).

Cultural Transmission and Change

Following Cavalli's earlier analogy with biological evolution, Cavalli and Feldman further developed the concepts of cultural mutation and cultural selection. On the cultural level a "mutation" is an innovation, a new idea or practice. Unlike biological mutations, these are not random but directed toward

2. Anthropologists were of no help here. Among them there are over one hundred different definitions of "culture" and little professional agreement as to what culture is.

the solution of perceived problems. A mutation can also arise through "copy error"—an older idea or practice is incorrectly transmitted or some error is made in how it is received. This latter is, of course, similar to random mutations in biological evolution. However it arises, a cultural mutation is either adopted or rejected, or adopted by some and rejected by others in a process of "cultural selection." Cultural selection is very different from natural selection in that it depends on human choices. What, then, determines whether or not or to what extent a cultural mutation is adopted? Cavalli and Feldman admit that this question moves us into the area of human motivations and drives, a complex and as yet not well understood subject. Humans may make choices toward individual "maximization of self satisfaction," but how this actually works is not known. They write: "The main limitations are the boundaries of our present knowledge on the functioning of steering mechanisms in our central nervous system" (1981:266).

What Cavalli and Feldman do assert, however, is that these human choices, which result in the adoption or rejection of cultural mutations, depend to some extent on how culture is transmitted. According to their theory, culture can be transmitted in a number of different ways, each of which may have different consequences for cultural evolution. There are three basic types of transmission: vertical, oblique, and horizontal. **Vertical transmission** refers to transmission from a parent or parents (biological or social) to a child. This type of transmission is the most comparable to genetic transmission, and like genetic transmission, it tends to be conservative. **Oblique transmission** is transmission to a child from persons who are not the child's parents but who are of the parental generation (or above it—for example, aunts, uncles, grandparents) or even persons unrelated to the child who are a generation or more above him or her and who serve as adult role models. Like vertical transmission, oblique transmission tends to be conservative although perhaps less so than vertical transmission. What is important about vertical and oblique transmission is that they are top-down and the recipients are relatively young. In Cavalli and Feldman's model, the introduction of oblique transmission was really made necessary by a requirement of the mathematical analysis. To simplify calculations, the time unit was set as one generation. It is possible to consider finer age differences, but this would complicate the mathematical study. For most purposes, oblique transmission can be ignored. Finally, **horizontal transmission** occurs between members of the same generation, or between or among persons where relative age and **kinship** relations are unimportant.

What remains essential is the distinction between vertical and horizontal transmission because they have very different properties. Evolution under vertical transmission is generally slower because it involves longer time intervals: generations. For this reason, it has a tendency to be as conservative as genetic evolution. The same transmitters and transmittees are involved in both genetic and vertical cultural evolution. On the other hand, in the horizontal mode, transmission usually takes much less than a generation, and hence horizontal cultural evolution can be very fast. With modern means of communication like the telephone, radio and television, and the Internet, horizontal transmission can be almost instantaneous.

A further factor affecting evolutionary rates is the number of individuals involved in each transmission act. In vertical transmission it is often just a one-to-one affair. In horizontal transmission the numbers of transmitters and transmittees can be different, and the rate of cultural evolution is profoundly affected by these numbers. The simplest form of horizontal transmission is one transmitter to one transmittee, and the latter becomes, with very little delay, a transmitter in another one-to-one situation.

One-to-one horizontal transmission follows an epidemiological model, working like the spread of a contagious disease through a population. News or gossip, for example, often spreads through a population through this type of one-to-one transmission. Other forms of horizontal transmission include many-to-one and one-to-many. In the first type, many persons are transmitting the same idea or practice to one individual. This mode of transmission is highly conservative; it tends to maintain the status quo through the application of what we conventionally understand as "social pressure" or "peer pressure." Because this many-to-one transmission is concerted, with every transmitter giving the same signal, this group effect provokes very powerful conformism, making evolution difficult and rare.

At the opposite extreme is a "one-to-many" horizontal transmission mode. An example would be a teacher transmitting knowledge and ideas to students. This mode can sometimes result in rapid culture change and can successfully spread more radical innovations. For example, a transmission of a new idea or practice to many individuals from a single powerful or charismatic person can be rapid. In many contemporary societies, the media acts as a powerful type of one-to-many cultural transmission that can effect rapid cultural change.

Cavalli and Feldman point out that these types of cultural transmission are not mutually exclusive in their operations. Rather, they see that in real-life situations, these modes can be operating at the same time, in concert or in conflict, and that many permutations among them are possible. What they seek in differentiating modes of transmission is to draw attention to the question, Who transmits what to whom (and when), in human cultures generally, or within a particular culture? They also seek to emphasize that some modes of transmission will tend to be conservative, discouraging change, and some will be potentially more radical and able to effect rapid culture change. To understand stable cultural differences, vertical transmission is particularly important. Indeed, to Cavalli, almost all characteristics that distinguish the different cultures existing in the world are the consequences of vertical transmission.

A key issue in cultural transmission is the age of the recipient of cultural information. As is well known, what humans learn earliest in life (for example their native language, patterns of personal hygiene) tends to be retained throughout life. Thus the idea that vertical (parent-to-child) transmission is conservative is directly related to the idea that a lot of what parents teach children is transmitted at the earliest phases in the children's life cycle. Cultures also differ in the extent to which parental instruction remains a primary mode of cultural transmission after infancy (as noted below).

Cavalli and Feldman then present their theory of cultural transmission in the form of sophisticated (and, to the mathematically challenged) unfathomable mathematical models. Here they apply mathematical models similar to those already developed in population genetics to vertical cultural transmission, assuming that cultural traits behave like genes, but with allowance, of course, for the fact that they have different and varied modes of transmission. Their models result in predictions of the probability of the spread of cultural mutations under various circumstances and conditions.

Cultural Evolution and Transmission is a purely theoretical work. Cavalli and Feldman did not base this work on any data, although they did give examples and illustrations of their points drawn from observations of everyday life or from the ethnographic literature. They also referred occasionally to the results of their study of Stanford University students, ongoing as the book was being written, and conducted in collaboration with sociologist Sandy Dornbusch. "This is the first time we actually collected data," said Feldman. This was a study of the similarities and differences in assorted characteristics (beliefs, habits, social activities, etc.) between a sample of Stanford University students, their parents, and friends. The study found greatest similarity overall between husbands and wives (parental pairs of the students), next between the students and their parents, and last, between the students and their friends. From this study it is not known how much of the strong husband-wife similarity was due to mate selection (people choosing spouses who have beliefs, habits, and so on similar to their own) or to a convergence of traits after marriage. But what was interesting in the study was the finding that there were strong similarities between students and their parents on issues of religion and politics. Little or no similarity was found between students and parents in many other habits and beliefs, ranging from items like the amount of salt used at meals to belief in superstitions. The strong similarity in the areas of religion and politics may be due to the fact that (at least in the United States) these are issues that parents tend to communicate to a child early in its life. Also interesting was the finding that two aspects of religion were strongly shared between students and mothers: frequency of prayer and choice of religion where the parents adhered to different religions.

Beginning in the late 1970s, one person, later to become a cultural anthropologist, came under the influence of Cavalli and, ultimately, helped to advance his and Feldman's theory of cultural evolution. This is Barry Hewlett, now professor of anthropology at Washington State University. Before meeting Cavalli, and before even entering a Ph.D. program in anthropology, Hewlett had traveled through Africa on his own and had encountered the Aka (hunter-gatherers), in whose culture he became very interested. Back in the United States, Hewlett wanted to meet Cavalli since at that time Cavalli was one of the few people to have written about these African tropical forest hunter-gatherers. When Hewlett walked into Cavalli's Stanford office one day, he was graciously and kindly received, even though "I was a nobody" and Cavalli was a prominent and busy professor. Hewlett was impressed that Cavalli took the time to talk with him and was "incredibly kind." They ended up spending hours talking about the Aka, during which Cavalli made suggestions about further research they could do together. And they did. Hewlett returned to Africa where he collected data on the Aka of the Central African Republic. Again back at Stanford, Hewlett reported that Cavalli patiently helped him to analyze data and write up results, and the two of them published together. Of this Hewlett said, "talk about mentoring!" To Hewlett, Cavalli is the person who has had the greatest influence on his own thinking and career. It was only later, after publishing with Cavalli, that Hewlett entered a Ph.D. program and eventually got his degree in anthropology, from the University of California at Santa Barbara (see fig. 4.5).

Figure 4.5 Cavalli (*center*) with Barry Hewlett (*right foreground*) and Hewlett's former student, Robert Moise (Vancouver, Washington, 2003). (From the authors' collection)

Later in life, Hewlett and Cavalli spent time together in the field with the Aka conducting research. Often Hewlett went first to conduct research on his own and to make arrangements for Cavalli to collect DNA samples later. Hewlett recalls the extraordinary energy Cavalli displayed in this challenging field setting. "Whenever he was there, it was an intense time." Hewlett also remembers that Cavalli considerably enhanced the quality of their field life. On his own, Hewlett ate the local hunted-gathered foods, but when Cavalli arrived, "We had pasta every day [and] good things to drink." Once in their camp Cavalli even concocted an elegant banana flambé!

In time Hewlett also became influenced by Cavalli's ideas on cultural transmission. Hewlett's major contribution has been to empirically test cultural transmission theory. He and Cavalli jointly applied this theory in a study of cultural learning among the Aka, publishing their work in 1986. This study suggested that the conservative nature of Aka culture is related to the observation that so much of the basic knowledge necessary for these peoples' hunting-gathering adaptation is vertically transmitted. A later study of several African cultures (where Cavalli and Hewlett collaborated with Italian colleagues C. R. Guglielmino and C. Viganotti) showed that among the cultures covered, the most conservative cultural characteristics have to do with kinship and family structure. This finding was confirmed in a subsequent study, published in *Current Anthropology* in 2002, which also suggested that kinship and family forms were primarily spread by demic diffusion in Africa.

Cultural and Biological Coevolution

As mentioned, Cavalli and Feldman were interested in developing an encompassing theory that combined biological with cultural evolution. They saw that cultural mutations and selection not only work in a way analogous to biological evolution but that these processes are interlinked. Their thinking can be diagrammed, as in figure 4.6. On the left we see natural selection operating on genetic mutations, resulting in human biological evolution. As part and parcel of evolution through natural selection, humans also developed their capacity for culture. Culture then became a primary means of human adaptation and a very powerful one, by making possible specifically directed, easily transmissible adaptations. Humans are at this point termed "primary organisms" by Cavalli and Feldman; their cultural products—technology, knowledge, beliefs, institutions, organizations, and so on—are in this scheme considered "secondary organisms." Moving to the right of the diagram, cultural selection then acts upon mutations occurring in these secondary organisms, driving cultural evolution.

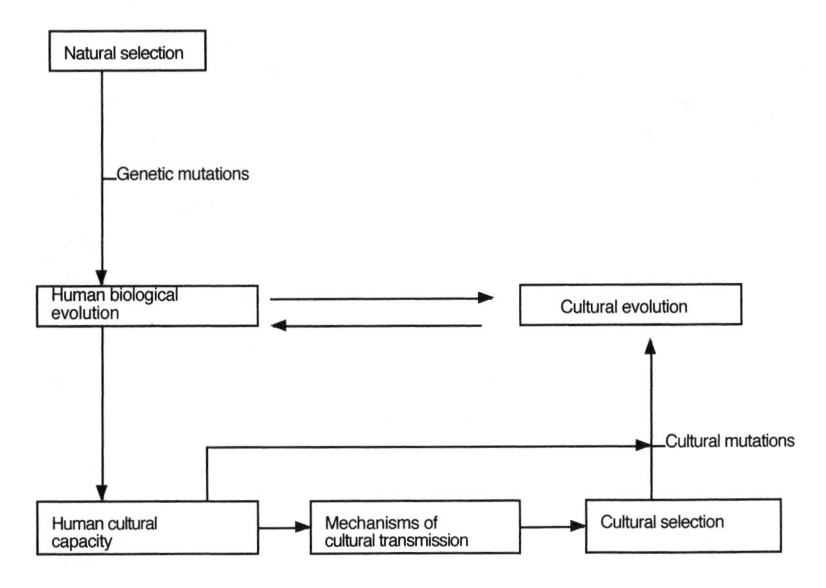

Figure 4.6 Interconnections between cultural and biological evolution.

Whatever evolves culturally is ultimately subject to Darwinian natural selection in the long run—secondary organisms that significantly decrease human biological fitness will not survive. Cavalli and Feldman call this natural selection a "second order selection." Before it operates, a "first order selection" takes place through cultural selection. Finally, cultural evolution and biological evolution are interrelated in humans and can affect one another. They may be in harmony or conflict and they may influence the direction of one another. Cavalli and Feldman wrote that "it might be argued that natural selection has been rather successful in insuring harmony between cultural and Darwinian selection" (1981:342–43), but note that this still leaves room for friction or conflict between them. Examples of persistent cultural practices that reduce fitness are the high fat and sugar diet in North America and female circumcision in some African societies. By contrast, lactose tolerance and the adoption of livestock herding are offered as examples of how Darwinian fitness and a cultural practice interact.

For Cavalli, cultural selection has paramount importance for determining the course of human events, even though it is under the ultimate control of natural selection. A wrong cultural decision may cost one's life or have strong negative effects in terms of natural selection. One relevant personal experience Cavalli had concerns the cultural rule of driving on the left (as in England) or on the right-hand side of the road (as in continental Europe). Cavalli bought his first car, a 13-year-old small Ford, when living in Cambridge in 1949, and went from there on holiday to Italy with his family, then composed of his wife Alba and their eldest son Matteo. The long trip to Europe and the return journey back across the Alps (there were no tunnels for cars at that time) were tiring. The Cavallis crossed the English Channel on the car ferry at night, arriving at Folkestone from Boulogne, France. It was foggy when they started driving back to Cambridge early that morning. At this time of the day, the road to London was mostly empty, but suddenly there emerged in front of Cavalli a big truck, the lights of which pierced the fog at the very last moment. He had inadvertently started traveling on the right-hand side of the road—the tiredness of the long drive the day before, the bad night, the early morning, the deserted road, and the dense fog had all combined to cause a terrible error. Swerving immediately to the left saved the life of the little family, but it was a very close call.

As nearly experienced by Cavalli, a wrong cultural decision can cost one's life, and in this case, the lives of family members. However, driving is but one activity which costs many lives, often because of someone's foolishness or error. Drugs are another obvious example. Further, Cavalli and Feldman in their book *Cultural Transmission and Evolution* examine a special case of a custom that almost destroyed the Fore tribe in the New Guinea highlands: necrophagy of close relatives. It is (or rather was, since this custom has now been eradicated) a duty of children to eat some brain tissue of their deceased parents, and close relatives in general, after cooking the cadavers with a wood fire placed on top of the burial mound. It also turned out that the disease called kuru (very similar if not identical to "mad cow" disease) was causing many deaths in the New Guinea highlands.

Because kuru was transmitted in families, it was initially described as a probable genetic disease. However, Carleton Gajdusek (who in 1976 was awarded the Nobel Prize for his discovery), having reproduced the disease in chimpanzees, showed that it was not genetic and could not be caused by a virus. Rather, we know now that kuru is caused by self-reproducing proteins called prions. Prions are heat-resistant, which did not make safe the Fore custom of cooking cadavers before eating them. Thus, kuru is an example of a deadly disease spread by a cultural practice. In other words, a local custom deeply affected the gene pool of some New Guinean tribes through culturally self-inflicted natural selection.

Interestingly, "mad cow" disease, also caused by prions, is now a global concern. The parallel with kuru is unmistakable because the disease was transmitted to cattle through the use of feed originating from processed meat from sheep that had died from a prion infection. As we know, this practice led to the destruction of countless heads of cattle in several countries, and much human suffering endured by those afflicted with the disease.

It can be said that natural selection will always be able to "discover" cultural errors, but it can mend them only by punishing the individuals or groups that made (or make) these errors. Because of the ultimate control of natural selection, cultural evolution may stay on the right course unless it causes total destruction. The example of nuclear weapons comes to mind.

Figure 4.6 shows connections between genetic mutation, natural selection, cultural mutation, and cultural selection. But it does not show another important source of cultural variation and change—namely, cultural drift. Cultural drift can come about through migration, when a migrating group carries with it a higher frequency of a cultural trait than was the case in the parent population. Cavalli has offered an interesting example: religiosity in the United States. As he wrote in *Genes, Peoples, and Languages*:

> [T]here is some indication that the U.S. culture is one of the most religious in the world. It is clear that there is a good reason. The religiosity of the U.S. population must come from a strong founders' effect, as shown by the fact that the major contribution to the U.S. culture was from English immigrants in the seventeenth century, who came mostly in search of freedom from religious persecution. American religiosity must be a case of cultural drift. (200–201)

Indeed, Puritanism, still found in many aspects of American culture, entered North America with English migrants. They were a small group but they prospered and influenced subsequent migrants.

Cultural drift can be very strong in cases of one-to-many transmissions, where the cultural transmitter responsible for proposing and/or spreading an innovation may have an enormous cultural impact on a multitude. For example, most religions were started by a single innovator. In such a situation, the founder influences his disciples through the one-to-many horizontal cultural transmission mode, but much of the growth in successive generations is determined by vertical cultural transmission because almost all children adopt their parents' religion. Additionally, truly successful religions are spread by conversion of adults (by proselytism), which is an example of horizontal transmission. In the case of religions, cultural drift is thus followed by both vertical and horizontal transmission.

Migration is also a very effective component of cultural transmission, both vertical and horizontal. Joseph Greenberg, the Stanford linguist, noted that linguistic innovations are often created by mothers when they come from a linguistic group different from that of their husbands. Seventy percent of the world's cultures are patrilocal, and in these cultures children learn to speak the father's language. However, in vertical transmission fashion, some of the mother's tongue can slip into the children's vocabulary or other aspects of their language. Nevertheless, most cultural innovations are spread among different populations by contact and learning, which is horizontal cultural transmission.

Migration has both genetic and cultural roles. Migration of individuals in the form of gene flow (see chapter 3) causes the introduction of genes into a population from outsiders, especially if it is sustained for a long time (even at a low rate). Migration tends to homogenize populations genetically, decreasing genetic differences previously built by drift or natural selection. To some extent migration may also homogenize cultures, though less markedly. But migration of a group, as in colonization of remote lands by people who settle voluntarily (as in the case of European settlers in all other continents), or by force (as in the Roman and Chinese empires, and for blacks brought as slaves to the Americas), also introduces new customs, as well as people. For example, voodoo was introduced into the Western hemisphere by slaves from West Africa.

Cavalli and Feldman's ideas inspired further work in biological or evolutionary anthropology in what became known as "**dual transmission**" or "gene-culture coevolution" theory, which addresses the transmission of both genes and culture and the interrelationships between genetic and cultural transmission. Notable here are William Durham, author of *Coevolution* (1991), and Robert Boyd and Peter Richerson, who together published *Culture and the Evolutionary Process* (1985). Both Boyd and Richerson in 1978–79 took a class at Stanford on cultural evolution taught by Cavalli and Feldman.

There are some differences between the dual transmission theories of these authors. For example, Barry Hewlett (2001) points out that Durham tends to see culture as more largely adaptive to the environment whereas Feldman and Cavalli, as well as Boyd and Richerson, show how cultural traits (or memes) can under certain circumstances develop a momentum of their own and play no adaptive role. But all of these dual transmission theories that include the transmission and influence of culture in human evolution should be sharply distinguished from the earlier **sociobiology** of E. O. Wilson and his followers. Sociobiology is also the application of Darwinian principles of natural selection to human beings; but unlike dual transmission theories, it posits direct, unmediated links between genes and human behavior, and it ignores the influence of human culture itself on the evolutionary process. To put this another way and in Cavalli's terms, sociobiology fails to distinguish vertical cultural transmission from genetic transmission.

The Response in Cultural Anthropology

Although used by Barry Hewlett and a few others, the Cavalli-Feldman theory of cultural evolution has not met with much acceptance in cultural anthropology. As Cavalli himself put it, "It was not so much rejected as ignored." There may be a number of reasons for this. One reason was discussed in chapter 1—that some anthropologists are concerned about the ethnocentric and racist history of the concept of cultural evolution. Cavalli has suggested another reason—that the theory was originally proposed through complicated mathematical models that anthropologists could not follow. As Feldman pointed out to us, cultural anthropologists are usually poorly trained in mathematics, and one of us (LS) can serve as a living example of that statement. Cavalli even went so far as to say that in retrospect it was not a good idea to include the mathematical models in the original presentation of the theory. The reason why they did so is that they wanted their theory to be rigorous. Ideally the book should have been followed by another, presenting the same ideas but for a larger audience. But Cavalli was by then working again on genetic evolution. Two subsequent popular science books by Cavalli (*The Great Human Diasporas* and *Genes, Peoples, and Languages*), however, contain a summary of the theory and its extension to linguistic evolution (see chapter 5).

Another possibility is that cultural anthropologists are uncomfortable with the idea of studying culture through examining discrete cultural traits. Cavalli and Feldman were aware of this reaction and in a later 1994 publication wrote:

> The theory of cultural transmission and evolution that we have developed over the past twenty years has been criticized by cultural anthropologists on the grounds that it focused on the "atoms" of culture, i.e., traits and their variants, and not on the "gestalt" of a culture. ("On the Complexity of Cultural Transmission and Evolution," 48)

This is true; cultural anthropologists prefer to see each culture contextually, not dismembered into traits or memes. They also consider that a cultural trait cannot be understood unless its meaning to the people who possess it is known. All this is not to say, however, that the Cavalli-Feldman theory is incompatible with the orientation of most cultural anthropologists, but only that it speaks to a broader level of analysis and of necessity uses particular methods of analysis. As mentioned earlier, Cavalli and Feldman's use of "culture traits" follows logically from their treatment of culture change as analogous to biological evolution, where "specific features of an organism, rather than the whole organism, have been the focus of attention" (1981:69). Cavalli notes that genetics started from the study of transmission of simple, sharply contrasted traits, and it has gone a long way in just one century. For Cavalli, it will be equally important, in order to understand culture, to study its transmission. This is because culture is the result of the accumulation of knowledge over millennia and longer. He once asked French anthropologist Claude Lévi-Strauss what he thought of cultural transmission, and the answer was : " It is too complicated to study." Cavalli feels this is not true; it is simply that nobody tried before.

Another reason for the lack of interest in the Cavalli-Feldman theory may have been that some anthropologists felt they had already considered, and by the 1970s largely dismissed, a somewhat similar idea known as the "early-learning hypothesis." As articulated by Edward Bruner in 1956, this hypothesis related culture change to socialization by specifying that what in a particular culture is "traditionally learned and internalized in infancy and early childhood tends to be most resistant to change." (197). This hypothesis was criticized for its narrow focus on discrete culture traits—a criticism also made against the Cavalli-Feldman proposal. The early-learning hypothesis was also criticized for its view of cultures as homogeneous entities that in one piece rejected or kept certain traits, ignoring the diversity within cultures among individuals and groups, especially in their reactions to situations of change.

It is important to note, however, that although the Cavalli-Feldman theory incorporates the idea that early learning may be relatively resistant to change, the real emphasis of this theory is on mode of transmission and especially the distinction between vertical and horizontal modes. What is intriguing about the theory is that it gives a great deal of strength and power to vertically transmitted culture and a great deal of human creativity and individual innovation to horizontal transmission. It was Barry Hewlett who in his 2001 article first noted the parallel between the Cavalli-Feldman vision of culture and the notions of Alfred Kroeber, who much earlier saw culture as reflecting human creativity (or even play) and as developing a direction of its own. This view challenges the alternative idea that much of human culture can be explained as an adaptation to a natural environment or as in some way promoting the survival and well-being of a group. By contrast, the Cavalli-Feldman theory offers students of culture change and persistence other possibilities—for example, that however a cultural belief or practice may have originated (whether or not it was initially adaptive in a specific environment), its current existence may be the result of its vertical mode of transmission and not its value as a mechanism of adaptation.

The lack of enthusiasm in cultural anthropology for the Cavalli-Feldman theory may be additionally that, just as Cavalli and Feldman were publishing their work in 1981, many cultural anthropologists were being influenced by the currents of postmodernism that were sweeping over the humanities in general (see chapter 1). In this movement, the position of *cultural relativism*, already established in cultural anthropology, gained strength and was taken to new levels. Cultural relativism asserts that each culture must be understood on its own terms, that is, with respect to its own concepts, ideas, and values. That much practically any cultural anthropologist would agree with. But some anthropologists took this notion a step further to assert that since each culture can only be understood on its own terms, we cannot do valid cross-cultural comparisons. Some saw anthropology itself as thoroughly enmeshed in specifically Western cultural concepts and values; therefore, translating a culture's own terms, concepts, and categories into the terms, concepts, and categories of anthropology was a distortion and, ultimately, an ethnocentric imposition of external, culturally biased meanings onto local meanings. These intellectual currents might have been reformative, encouraging anthropologists to be aware of biases in their own thinking and so to improve their arguments; but a real fear was that they seemed to be headed on a path of disciplinary self-destruction.

Many postmodernists emphasize that all knowledge is conditioned by observer bias. What we "discover" out in the world is actually based on our position in the world; our angle of vision stems from our cultural background, social class, personal interests, and so on. Objective knowledge or truth is not possible; rather, humans only devise modes of thought or ways of knowing that are relative to one another, none having superior truth value. This position is sometimes termed *epistemological relativity*. As pointed out by anthropologist Philip Salzman, postmodernists combine epistemological relativity with moral advocacy, holding, for example, that what anthropologists should do is not aim for "truths" about culture, but should instead give the oppressed people of the world (women, the poor, certain ethnic groups, etc.) the opportunity to voice their own perspectives and concerns and to help end their oppression by others.

Some cultural anthropologists influenced by postmodernism abandoned the project of anthropology altogether and confined their writings to critiques of the cultural biases in the work of other social scientists. Other anthropologists who continued research and interpretation of culture moved closer to an attempt to understand a culture from within, in the process scrutinizing and making explicit their own cultural biases. Accordingly, they tended to veer away from grand, cross-cultural theorizing and even further away from any integration of their work with biological or other sciences.

Finally, as discussed in chapter 1, some postmodernists questioned the privileged position of science itself as a superior mode of knowledge. Indeed, many—even those who otherwise shunned all social theory—saw power struggles everywhere, especially along the axes of men over women, white people over people of color, wealthy classes over poorer ones, anthropologists over their "subjects," scientists over nonscientists. Those at the top were seen to engage in "hegemonic discourses" to maintain their power. Thus any kind of scientific anthropology, like science itself, hegemonically promoted inequality and injustice.

With all this going on in cultural anthropology, it is perhaps not surprising that the broad, mathematical, scientific, and macro-level Cavalli-Feldman proposal was ignored. To some anthropologists it might have looked like a program for Martians to study Earthlings. More recent events in anthropology may, however, give it, and its offspring theories, another chance. Many within the field saw the rising tide of postmodernism as destructive and becoming too extreme. They began to resist and, while taking note of postmodern critiques of academic cultural biases, held to the notion that a scientific study of human culture is both possible and legitimate, that it is not necessarily "evil" or hegemonic. Within anthropology the tension between those who seek a scientific study of human culture and those strongly influenced by postmodernism mounted and began to tear the profession apart. Once again, Stanford University was at the center of things. In 1998–99 the former Department of Anthropology split into two parts, one called the Department of Anthropological Sciences and the other called the Department of Cultural Studies. Then, in 2002, a group of anthropologists founded a new group, the Society for Anthropological Sciences. This group began with its founders' claim that several proposed symposia for the 2002 annual meetings of the American Anthropology Association (AAA) were rejected because they were scientific. What they perceived as an antiscientific bias within the profession led them to hold separate but parallel professional meetings of their own in a hotel across the street from the main AAA conference meetings (in New Orleans in November 2002). This is a recent event and only time will tell what will happen to the profession of anthropology in the United States and potentially elsewhere.

Some cultural anthropologists are likely to affiliate with the Society for Anthropological Sciences; and within this group several ideas and theories of culture formerly suppressed may have a new hearing. Here, the Cavalli-Feldman proposal, along with other ideas, may come to be seen as having a number of advantages for scientifically oriented cultural anthropologists. For one, it allows at least the investigation of a theoretical integration of biology and culture, previously an almost forbidden union in cultural anthropology. Second, the Cavalli-Feldman proposition is a theory of culture change, and as such should be of interest within cultural anthropology, particularly in terms of how it might fit or conflict with other theories.[3] Third, the theory

---

[3]. Currently in cultural anthropology an approach known as *cultural materialism* (developed mostly by Marvin Harris) addresses cultural change. This theory, with its roots in the work of Karl Marx, Leslie White, and Julian Steward, looks first to the material conditions of human life in an attempt to understand cultural differences and similarities. Cultural materialism holds that "much of cultural evolution has resulted from the gradual accumulation of useful traits through a process of trial and error" (Harris 1997:426).

offers testable hypotheses. Fourth, this theory is compatible with some current trends in cultural anthropology that most likely will continue in the "scientific" wing. For example, contemporary cultural anthropology recognizes individual variation (sometimes called *multivocality*) within any culture. People within a culture do not all speak with one voice; rather, women may have different beliefs and practices than men, people's ideas may vary with age, ethnicity, social class, and so forth. Another current emphasis is on individuals as actors making choices, rather than slavishly absorbing whatever their culture tells them. This too (as discussed by Hewlett in 2001) is compatible with the Cavalli-Feldman theory of individuals as cultural innovators and choice-makers guiding cultural change.

A final advantage of the Cavalli-Feldman proposal and related ideas is that it opens the possibility for cultural anthropology to be cumulative in its theory as well as its substance. So far, cultural anthropology has tended to follow intellectual fashions (many stemming from France), continually discarding what is old and trying on for a while what is new. Theories are never disproved but, rather, deemed to be "flawed" and "inadequate," after which they are abandoned. Any movement away from this kind of theoretical wheel-spinning in cultural anthropology might be welcomed by many. The challenge will be to combine a more macro-level, broad theory such as Cavalli and Feldman's with the more close-range perspective of the cultural anthropologist, a perspective that is also sensitive to cultural context and to observers' own cultural biases. Already contributing to this endeavor is Barry Hewlett (as we discuss in the chapter 8).

Chapter 5

Genes, Languages, and Human Prehistory (1970–)

By now we have seen a few recurrent themes in Cavalli's career: an integration of knowledge from different fields and openness to collaboration with others, an active imagination and a willingness to follow it into new directions. As Marcus Feldman put it to us, "[Cavalli] has no boundaries on the scope of his imagination. . . . Working with Luca opens up intellectual pursuits that are not normally viable." Another colleague remarked, "He hasn't just focused on a tiny little corner. I have always found that extremely inspiring." Cavalli's imagination and interdisciplinary abilities come to the fore in an important phase of his career that concerns his contributions to world prehistory, contributions that integrate genes, languages, and human migrations.

As evidence of his imaginative insights, Cavalli was among the very first to consider that genes of living persons could carry traces of the past, or that genetics could be an instrument in the unraveling of human evolution and prehistory (see chapter 3). By now there are many examples outside of Cavalli's own particular work that show how genetic data can decisively settle some evolutionary issues. An early one was the comparison of human genes with those of chimpanzees and other primates, carried out in the 1970s and 1980s. These comparisons showed that humans are most closely related to chimpanzees and that in fact we share 98 percent of our genes with this species. Furthermore, using the molecular clock with a constant rate of genetic evolution, it was calculated that humans and chimps began to diverge about 5–7 million years ago, a date supported by the fossil record. A very recent example of the use of genetic data in the study of evolution is the analysis of DNA from Neanderthal remains, which strongly suggests that Neanderthals did not interbreed with *Homo sapiens* or make genetic contributions to our species.

In Cavalli's work, genetic data were collected and analyzed to address the question of where modern humans arose and how they came to populate the earth. His work has been massive, covering in some detail the spread of humans throughout the globe, combining genetic data with data from linguistics and archaeology. His contributions to human prehistory have been published in a number of books and articles, most thoroughly in his major book, *The History and Geography of Human Genes* (written with P. Menozzi and A. Piazza, 1994).

In this chapter we highlight Cavalli's contributions to prehistory by focusing on how his work has shed light on particular issues or questions. These include the debate over where and how humans evolved (the multiregional versus uni-regional—or, ``Out of Africa"—hypotheses), the peopling of the New World, the origin and spread of Indo-European languages, and the question of whether there was one single early human language from which all modern languages evolved. In addition, throughout all his contributions to these issues is an underlying one that concerns the relationship between linguistic and genetic evolution. Before moving to these topics, however, we will briefly review just what it was that Cavalli did, or how he applied genetic data to human prehistory.

Genes and Prehistory

As we saw in chapter 3, it all began in 1960, when Cavalli collaborated with Anthony Edwards, a student of R. A. Fisher. Cavalli had invited Edwards, an expert in population genetics and statistics, to join him at Pavia for this work. At first they only worked with blood groups and had data from only a few populations (fifteen in all, three per continent; later they worked with more populations and more genes). They obtained from the literature the gene frequencies of the blood groups of the populations and then calculated the genetic distances between them, as explained in chapter 3. The only error in the first tree they drew using blood groups (which grouped African and European populations close together and gave a major distance between European and Asian populations) was later corrected with the addition of more genes and the realization that the root of the tree was actually located in Africa (for the updated, corrected version of populationrelationships, see figure 6.1). Initially, due to the lack of a method for fitting the root of the tree, Cavalli and Edwards placed it arbitrarily in the middle of their diagram. As we will see later in this chapter, it was Allan Wilson's 1987 "Out of Africa" hypothesis, based on mitochondrial DNA, that gave the first solid hint that modern humans originated in Africa (Cann, Stoneking, and Wilson 1987). It should be noted that others, including Cavalli and Bodmer, had suggested earlier that modern humans may have first appeared in Africa. It is Wilson's laboratory, however, that provided the first detailed data supporting this idea.

We have seen in previous chapters that with data on population gene frequencies, Cavalli and his coworkers were able to construct principal component maps, showing distributions of genes in space. With this same data, and with a calculated rate of genetic change, they could also begin construction of phylogenetic "trees" showing the historical connections between populations (chapters 1 and 3). Tree construction is a complicated matter, the details of which and some of the problems with which will not be discussed here. We offer instead a sample tree (fig. 5.1) based on

1. African
2. North African and West Asian
3. European
4. Amerind
5. Arctic Northeast Asian
6. Northeast Asian
7. Southeast Asian
8. Pacific Islander
9. New Guinean and Australian

Figure 5.1 A simplified phylogenetic tree of the human family. Modern humans (ancestor) first appeared in sub-Saharan Africa between about 150,000 and 100,000 years ago. Around 50,000 years ago some of them migrated out of Africa and differentiated into the groups indicated. The depth of the "well" linking African populations to the ancestor indicates that they appeared first. According to this tree, modern Australians and New Guineans differentiated next. Modern Europeans and other Caucasoids, as well as Native Americans, appeared last. (Redrawn from L. L. Cavalli-Sforza and M. W. Feldman. 2003. The application of molecular genetic approaches to the study of human evolution. *Nature Genetics* 33:266–75)

the most recent data published in the literature. This tree shows how these populations have genetically descended from an ancestral population and branched off from one another, with an initial split separating African populations from the rest.

Over the years, Cavalli and his coworkers have considerably refined their trees, have added hundreds of genes to their data bank, and have adopted new methods of analysis. They have moved from the analysis of blood groups to other blood proteins, as well as to Walter Bodmer's much-studied human leukocyte antigens (HLAs), and DNA. More recently and particularly important in these studies has been the use of so-called **restriction enzymes** for the analysis of DNA polymorphisms. We give here a brief description of how this technique works because it is very important in the historical context of the study of human origins. In fact, Allan Wison's group generated the "Out of Africa" hypothesis entirely based on this technology.

Many proteins catalyze the numerous metabolic reactions that take place inside living cells. Catalytic proteins are called **enzymes**. A special class of bacterial enzymes, which act on DNA, are called *restriction enzymes*. The function of these proteins is literally to cut DNA into small-sized fragments. What is more, there exist many such enzymes in the bacterial world, and all these different enzymes cleave DNA at different points, depending on the sequence present in the DNA to be cut. Thus, different restriction enzymes have the ability to generate fragments of different lengths when they cut the same DNA molecule. These fragments can be separated by gel electrophoresis (a technique measuring the rate at which pieces of DNA move in an electric field) and their lengths measured.

Consider two DNA molecules isolated from two different people. Further assume that these two DNA molecules contain slightly different sequences because the two human DNA donors are variants of one another at a particular spot in their genome. In other words, their genome is polymorphic at that spot. The DNA sequence of individual #1 could be, for example, with only one of the two DNA strands shown for simplicity,

ATGCCTTGGAATCT**T**GGCATTAGGCTGGCTTGGAAACC

whereas the DNA sequence for individual #2 reads at the same spot

ATGCCTTGGAATCT**A**GGCATTAGGCTGGCTTGGAAACC

We see that #1 has a T base whereas #2 has an A (both shown in bold above), all other bases being identical. This is a very simple example of DNA polymorphism, taking place at the level of a single base. It is called *single* **nucleotide** *polymorphism* or SNP (pronounced "snip"). Consider now a restriction enzyme that cuts DNA at the level of a TTGG sequence (underlined). DNA from individual #1 will be cut by the enzyme because three precise TTGG blocks of bases are present. After cutting, four DNA fragments will be observed. These four fragments can be separated and their lengths measured. Now, what will happen if the same enzyme is allowed to cut DNA from #2? This DNA will generate only three fragments, because the second TTGG sequence is *not* present, having been mutated to a TAGG sequence, which is not cut by the enzyme. Thus, DNA from #2 will yield a three-fragment pattern, which is easily distinguishable by gel electrophoresis from the four-fragment pattern generated by DNA #1.

This technique is called *restriction fragment length polymorphism analysis,* abbreviated as RFLP analysis. This type of analysis is used extensively in the study of bacterial, fungal, plant, and animal (including human) evolution. For future reference, a stretch of DNA where the base sequence differs slightly from individual to individual, as in examples #1 and #2 above, is called a **haplotype** when this base-sequence difference is closely linked to other unique sequences located on the same chromosome. The term *haplotype* was invented by the Italian geneticist Ruggero Ceppellini, one of Cavalli's good friends. The simple example that follows further clarifies the concept. As we know, humans carry two sets of chromosomes. Let us now imagine three genes or DNA segments that we will call A, B, and C. Let us further imagine that these genes or DNA segments also exist in the form of variants a, b, and c. Suppose now that an individual possesses one copy of each variant. Using a bar to separate individual chromosomes, we find that the six variants can exist in four possible configurations: ABC / abc, Abc / abC, AbC / aBc, and aBC / Abc. If the three genes or DNA segments are located close to each other on the chromosomes, there will be little chance that recombination (see chapter 2) between them will occur. In the absence of recombination, we can thus distinguish eight possible haplotypes—ABC, Abc, AbC, aBC, abC, aBc, Abc, and abc—that will be passed on to progeny without shuffling. Thus, individuals, and populations, can be characterized as having a given haplotype or another. The notion of haplotype is used also in the classification of populations into different groups which have a recent common ancestor (as determined by Y chromosome DNA genealogy), in which case the word *haplogroup* is preferred. A similar electrophoresis technique (but not involving restriction enzymes) can be applied to the case of proteins. Today, RFLP analysis has been almost totally supplanted by the **polymerase chain reaction** (PCR) DNA amplification technique, often combined with DNA sequencing (see next chapter).

These and other equally complex methods of analysis, along with statistical techniques, lie behind Cavalli's use of genetic data to determine how closely or distantly populations are related to one another and to plot prehistoric human migrations. His data, providing genetic evidence of who was where when, has shed considerable light on some key issues and debates in the study of human origins and diversification. We now turn to these issues.

Multiregional Versus "Out of Africa" Hypotheses of Human Origins

One very important debate to which Cavalli's work has contributed is that between those who support the multiregional hypothesis of human origins and those who favor the "Out of Africa" theory. The multiregional hypothesis, as put forth by M.H. Wolpoff and others, states that modern humans evolved from *H. erectus* but independently in different regions of the Old World. Thus, the Chinese people are descended from the *H. erectus* of China (formerly called Peking Man); Indonesians evolved from the *H. erectus* there (Java Man), Europeans have European *erectuses* as their particular ancestors, and so on. Multiregionalism thus holds that modern humans do not share a common *H. sapiens* ancestor. This hypothesis amounts to what is known as parallel evolution, where two or more lines evolve separately but in a similar way so that the process results in the formation of one species or in the development of two or more closely related species. Old and New World monkeys are an example of parallel evolution. In human evolution, multiregionalists point to fossil evidence to support their theory, especially fossils of the skull. They claim there are clear similarities between, for example, *H. erectus* fossils of China and modern Chinese people.

Although not associated with racism today, an earlier variant of the multiregional hypothesis did carry racist overtones. A major proponent of this variant was anthropologist Carleton Coon, who published his ideas in the 1960s. Coon held that different lines of evolving humans crossed what he called the "*sapiens* threshold" at different times, producing the different racial groups into which he saw the human species clearly subdivided. Then, those lines that crossed the *sapiens* threshold earliest, and have thus been in the "*sapiens* state" the longest, are more advanced than those lines that crossed it later. Needless to say, Coon postulated that white people (Caucasians) crossed the *sapiens* threshold first and black people last! The reverse turns out to be closer to the truth. As we will see, African populations are the oldest of all human groups, so one could say that they have been in a "*sapiens* state" longer than other populations that descended and diversified from them.

The "Out of Africa" hypothesis holds, by contrast, that modern humans originated in Africa alone (evolving from African *H. erectus* and *Australopithecus* before it) and then spread out from Africa to the rest of the world. In this theory, all contemporary humans do have a common *H. sapiens* ancestor who once lived in Africa. In 1987, while Cavalli at Stanford was himself analyzing mitochondrial data to answer questions about the origin and spread of modern humans, a team at the University of California, Berkeley, published in *Nature* stunning mitochondrial genetic findings that strongly suggested that Africa was truly the homeland of modern humanity. This team was led by evolutionary biologist Allan Wilson. They analyzed the mitochondrial DNA of 135 contemporary individuals from different regions of the globe and determined the number of mutations in mitochondrial DNA between them. As we saw in chapter 1, a population with a higher frequency of mutations is considered ancestral to a population with a lower frequency of mutations in its DNA. This is because older populations have had more time to accumulate mutations than have newer populations. At the individual level, one can say that if two persons carry more mutations than two other persons, then the first two individuals are of older ancestry than the other two. With this, the Wilson team was able to construct a tree showing how closely or how distantly these 135 individuals were related through sharing common ancestors in the remote past. This tree showed that Africans diverged from an ancestral population earlier than had other groups. They also calculated a rate at which mitochondrial mutations accumulate, using the number of such mutations from chimpanzees. With this rate they could then date the point at which any two individuals shared a common female ancestor, since mitochondrial DNA is uniquely transmitted from mothers to their offspring.

Wilson's team also suggested that at the furthest point back in time, all humans shared one ancestress: a woman whose original mitochondrial DNA was transmitted (with changes in base sequence) through the female line to all modern humans. Among all human populations, African ones are the most genetically diverse. This means that they are the oldest. Thus, our ancestral woman, from whom all modern humans have descended, must have lived in Africa. She then became know as "African Eve," and we will meet her again in the next chapter. Wilson calculated a date of 190,000 years ago for African Eve, with approximate error limits of 150,000 and 300,000 years ago. We will see later the present state of knowledge on this issue. As Cavalli told us, Wilson was courageous to publish such revolutionary results implying an African origin of modern *H. sapiens* and the existence of an African "Eve."

There were criticisms of Wilson's work, and as a result the "Out of Africa" hypothesis was not immediately accepted. These criticisms ranged from the fact that African-Americans (and not Africans from Africa) were used in the study, to the existence of alternative phylogenetic trees, based on the same data, that did not support the "Out of Africa" model. Following some refinement, it is now accepted by a slight majority of researchers, thanks in part to Cavalli's work, which has supported the "Out of Africa" theory in at least three important ways. First, Cavalli claims that, from a genetic point of view, all modern humans are simply too similar for there to have been a case of parallel evolution over such a long period of time—that is, spanning the time between *H. erectus* and the appearance of modern humanity. Second, Cavalli's data (like that of Wilson and others) show that Africans are more genetically diverse than other populations, suggesting that theirs is the oldest population. Third, Cavalli's genetic data (after his and Edward's early work on blood groups) have consistently, and through many different methods and techniques, confirmed a major expansion out of Africa to South and Southeast Asia (most probably along the southern coast of Asia). From there populations moved to Oceania, with perhaps slightly later a migration to Central Asia, from where there was an expansion in all directions: Europe, Siberia, North Asia (and from there to the Americas), East Asia, and from there again to Southeast Asia and one more time to Oceania and to America.

What Cavalli and his colleagues did was to compare the genetic distances between various populations with the archaeological dates for first settlement on the various continents. For this they used polymorphic proteins (also called "classical" polymorphisms; see chapter 6). They got a nice match, as seen in table 5.1.

The genetic distances in this particular table were based on 42 populations, grouped into 9 categories with 110 genes analyzed. Here, the largest genetic distance (that between Africa and the rest of the world) is set at 100 so that all the other genetic distances could be recalculated as percentages. This finding of the greatest genetic distance to be between Africa and the rest of the world confirmed the earlier work of geneticists Masatoshi Nei and Arun Roychoud-

Table 5.1 Alignment of Genetic Distances with Archaeological Dates for Populations

Location / Population	Genetic Distance	Archaeological Dates (in thousands of years ago)
Africa and rest of world	100	100
SE Asia and Australia	62	55–60
Australia and Europe	48	35–40
NE Asia and America	30	15–35

hury. Although the archaeological dates in this table are imprecise, they could still be used to calculate a rate at which genetic distances increase with time after populations separate.

With these dates—and considering that greater genetic distances would mean longer times since separation—Cavalli suggested the following pattern of expansion, from earliest to most recent:

Africa ⟶ Asia
Asia ⟶ Australia
Asia ⟶ Europe
Asia ⟶ America

This is the human migratory path shown earlier in figure 1.3. The details of Cavalli's work are presented in the voluminous work *The History and Geography of Human Genes* (1994), which provides a careful discussion of his methods and a detailed coverage of human prehistory both globally and regionally. It also carefully matches genetic with archaeological and linguistic data. Very consistent with Cavalli's style, this work also mentions the limitations of the data and methods, the need for further research in particular areas, and the viability of some alternative interpretations of particular data. It is a marvelous summation of decades of work, presented in the best scientific and objective spirit.

This important work was, unfortunately, temporarily interrupted in 1991 when Cavalli suffered a heart attack followed by by-pass surgery. He recovered, and after 1993 he developed research with newly discovered genetic markers—microsatellites at first and, later, Y chromosome polymorphisms—that metamorphosed the field (see chapter 6).

The Peopling of the New World

Cavalli's research sheds new light on the most recent major human expansion—that of modern humans from Northeast Asia to the New World. Few have ever disputed that such a migration took place, with humans entering the New World in the north and then migrating south. Even multiregionalists support this scenario; since no *H. erectus* remains (or any other pre-*sapiens* remains) have ever been found in the New World, modern humans could not have evolved there but had to have come from somewhere else. And it is now fairly well agreed that East Asian people arrived in the New World by crossing an ancient land bridge (**Beringia**, now the modern Bering Strait) that connected Siberia with Alaska, thousands of years ago.[1] Another possibility is that migrants used boats along a coastal route to reach the New World.

While there is almost general agreement that it was East Asians who entered the New World, the question of *when* is surrounded by intense debate in archaeology. This is significant because without a firm time frame for entry into the New World, many cultural developments in the Americas cannot be properly interpreted. One problem is that the oldest artifactual remains have been reported from the South, almost suggesting a reverse migration from South to North. Further, these early southern dates are disputed. The upshot of all this is that archaeologists' dates for entry into the Americas vary from 14,000 to 35,000 years ago, an unusually wide spread of estimation in the archaeological record for these times. The oldest human skeletal remains in the New World with undisputed dates are found in Montana. But these have a fairly recent date of 10,600 years ago.

Another problem has been that some evidence for very early human occupation in America has been thoroughly discredited. For example, a claim of skeletal remains in California with a very early date of 50,000 yeas ago was shown to be incorrect when refined carbon-14 dating methods were applied. As a result of these and other dating mishaps, skepticism continues to surround any other evidence of early occupation. Two such cases are now hotly debated. One occurs in a site in Monte Verde, in southern Chile. This is a well-preserved site generally dated between 13,000 to 12,500 years ago; however, beneath this settlement archaeologists have found crude flaked tools that were dated at 33,000 years ago by T. D. Dillehay and M. B. Collins. Another case is a site in Brazil with crude quartz tools dated at 31,500 years ago by N. Guidon and G. Delbrias. Both of these dates have been criticized by T. F. Lynch. But if these early dates from South America are correct, then the entry of humans into the New World would need to be pushed back even further.

For a long time, the mainstream archaeological position on the peopling of the New World posited an entry through Beringia at about 12,000 years ago by the so-called Clovis people. The name *Clovis* comes from an archaeological site at Clovis, New Mexico, where in 1932 a number of fluted stone spear points were discovered. Later these distinctive points were found at

---

1. Earlier, archaeologists spoke about an ice-free corridor in the region through which humans could have migrated into the New World. But some archaeologists pointed out that such a corridor would not have provided a hospitable environment for population migration. It would not have, for example, supported trees. Since the corridor is hundreds of miles long and could not have been crossed quickly, this raises the question of what those who crossed it used for food during their passage.

hundreds of other sites in North America, and these fluted spearheads became known as **Clovis points**. The earliest Clovis points are dated at 11,800 years ago.

The finding of pre-Clovis sites in both North and South America, with dates of artifacts far earlier than 11,800 years ago has challenged the orthodox position to the point where E. Marshall describes a "pre-Clovis movement" and others talk about "the Clovis wars" between anthropologists. Those adhering to the earlier mainstream position that the New World was first settled by the Clovis people still continue to criticize the early dates with charges of sample contamination and other problems in accurately dating the samples. But today most archaeologists consider that the first people to enter the Americas were not Clovis people but precursors and that the Clovis tool kit was a later development.

What does the genetic evidence tell us? Cavalli's work shows a close genetic similarity between Native Americans and East Asians, which is no surprise to anyone. But in addition to this, his genetic evidence assigns a date of expansion from Asia into the Americas at 32,000 years ago, lending considerable support to those archaeological estimations that are on the early side. This was computed by taking the genetic distance between Amerindians and East Asians (6.6, as calculated from blood groups and protein polymorphisms) and calculating how long it would take to reach that figure, based on the other human expansions with known archaeological dates, as seen in table 5.1. However, this genetic distance assumes a constant rate of evolution. If the first settlers in America were few, as seems likely, then the genetic distance between Amerindians and East Asians would seem to be higher, and the calculated date of separation might be too early. Only further archaeological work will definitively answer the question of when Asian populations crossed into America, but Cavalli's results do suggest that the claims for early occupation need to be taken seriously.

Cavalli's work on the peopling of the New World also carries implications for studies of Native American languages. This calls up yet another debate. Most American linguists claim that the Native American languages (including both North and South America) are quite diverse; they posit that they span some two hundred different linguistic categories or families. Opposed to this has been the work of Joseph Greenberg, who classified Native American languages into only three families: Na Dene, Eskimo-Aleut, and Amerind. Cavalli's work also shows three genetically distinct categories that correspond to these three language groups. Further support for Greenberg has come from some archaeology that suggests three major migrations into America. The first was apparently the Amerinds, who spread everywhere; next were the Na Dene and Eskimo groups, which remained in the North. While some genetic evidence supports this, other genetic evidence does not. We will revisit the whole question of language classification in this chapter's final section.

The Origin and Spread of Indo-European Languages

In chapter 1 we introduced the Indo-European family of languages. This family consists of a number of branches: Germanic (which includes English), Italic (Latin and many languages descended from Latin such as French, Italian, and Spanish), Balto-Slavic (which includes languages such as Polish and Russian), Celtic, Greek, and Indo-Iranian (including Hindi, Urdu, Persian, and many others). Although Indo-European is the most thoroughly researched of all language families, considerable debate has surrounded the question of where and how long ago it originated and how it spread. Most linguists came to agree with archaeologist Marija Gimbutas, who proposed the area north of the Black Sea in a steppe region of what used to be the Soviet Union as the place of origin. She further proposed a date of 5,000 to 5,500 years ago for the antiquity of Indo-European, based on her study of archaeological remains in this region. These remains are associated with a culture known as **Kurgan**, noted for its burial mounds, a pastoral economy, domestication of the horse, and war chariots. Linguists and historians agree that one expansion of the Kurgan people went westward into Europe beginning about 5,500 years ago. Another branch spread southeastward over Iran, Afghanistan, Pakistan, and India about 3,500 years ago. These people became the Indo-Aryans, who settled over South Asia, pushing Dravidian speakers further south and possibly furthering the demise of the ancient Indus Valley civilization, or the people of Mohenjo-Daro and Harappa in what is now Pakistan.

As mentioned in the previous chapter, archaeologist Colin Renfrew developed an alternative theory. First, Renfrew gave a much earlier date of about 10,000 years ago for Proto-Indo-European. Second, he proposed the area of Anatolia, Turkey, as the place of origin of this language family. This place of origin is also supported by some Russian linguists such as A. M. Dolgopolsky. And third, Renfrew suggested that these Indo-European people were agriculturists who spread their language along with agriculture as they expanded east toward Europe, establishing farming communities first in Greece by 8,500 years ago and spreading elsewhere in Europe from there.[2] As we saw in chapter

[2] Renfrew also claimed that Gimbutas's dates for expansions from southern Russia, ranging from 5,500 to 3,000 years ago, are much too early and that the domestication of the horse was not as early as believed by Gimbutas and others.

4, this idea supports Cavalli's "demic diffusion" hypothesis, or the idea that farming spread into Europe through the migration of Neolithic farmers to Europe from the Middle East.

Some linguists object to Renfrew's theory, saying it is not supported by reconstructions of the Proto-Indo-European (PIE) language. That language, they point out, contained words for items such as *wheel* and *yoke,* which the Anatolian farmers of 10,000 years ago lacked. There is no evidence of wheels, for example, before about 5,500 years ago. Thus, the Indo-European languages could have only begun to diverge from their parent language, PIE, after items like these were in use and terms for them present in PIE. But Renfrew counters that root words for such items could have existed in PIE much earlier if we accept that their meanings could have been slightly different. For example, the PIE word for what became known as *wheel* in various Indo-European languages may simply have meant "to turn" in the PIE of 10,000 years ago in Turkey.

While Cavalli's genetic data supports Renfrew, what is even more interesting is his early suggestion (now followed by some others) that the theories of Renfrew and Gimbutas are not necessarily contradictory, and in fact probably complementary. This idea he bases on his principal component maps of Europe (specifically the first—see fig. 4.4—and the third principal components), which suggest that two major migrations may be involved in the spread of Indo-European languages into Europe. He agrees with Renfrew on Anatolia as the place of origin of Indo-European. From here a primary expansion of Neolithic Proto-Indo-European speakers, beginning about 9,000 to 10,000 years ago, went north to the steppe region of the Ukraine and west into Europe, bringing early forms of languages like Albanian, Greek, and Armenian. In the steppe region, cold and not conducive to agriculture, the Neolithics dropped farming as an exclusive mode of subsistence and developed pastoralism (seminomadic animal herding), a subsistence mode far more adaptable in the rugged steppes. Then later, around 5,000 to 5,500 years ago a second set of expansions took place from this steppe area southwest to Europe, as originally proposed by Gimbutas. These waves brought in the Celtic, Italic, Germanic, and Balto-Slavic language branches, possibly in that order.

Cavalli has consistently maintained that new technological innovations trigger major human expansions. In the case of the primary expansion of Indo-European speakers into Europe, the innovation was agriculture. For the secondary expansion of Indo-European speakers into Europe from the Ukraine area, the technological innovations were the development of bronze, the domestication of the horse, and the use of chariots, which gave these pastoralists mobility and military advantage. It should be noted that Renfrew insists that Gumbutas' and others' suggestion that horse mounting was as old as 5,000 years ago is archaeologically unjustified.

This idea of two important expansions of Indo-European speakers into Europe (the first speaking a kind of pre-Proto-Indo-European and the second speaking a Proto-Indo-European language) is intriguing. It is compatible with Cavalli's genetic data in the form of principal component analysis and, if correct, it would reconcile a long-debated issue, showing how two contested theories, both supported by a lot of good evidence, are in fact both right. Additional archaeological research could further test this proposal.

A Single Ancestral Human Language

There are over 6,000 languages spoken in the world. As we have seen with the Indo-European example, many of these languages are historically related to one another, and by using various techniques, linguists can demonstrate these relationships and so group languages into larger family units. These language classifications differ among linguists. Merritt Ruhlen, whose work Cavalli has followed, has developed a classification of seventeen language families. Figure 5.2 shows sixteen these families and their distribution on a world map. There are also some linguistic "isolates," or languages that seem to lack affinity with any current language family.

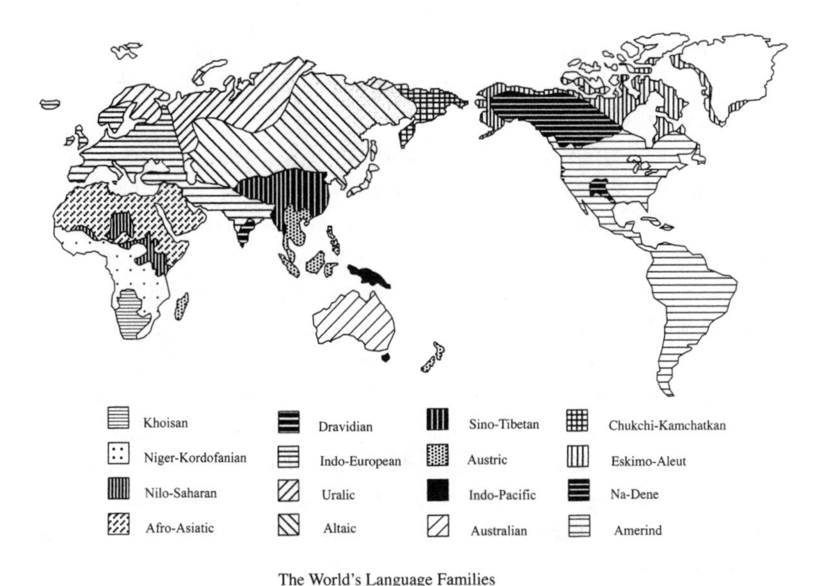

Figure 5.2 Merritt Ruhlen's language families distributed on a world map. (Courtesy of Dr. Merritt Ruhlen)

Linguistics has from its beginning been plagued with taxonomic disputes. As with many fields that use a **taxonomy**, there are the "lumpers" and "splitters." Lumpers focus on similarities they see between languages and tend to group them together into a few categories, implying historical relationships; splitters focus on the differences between languages and so separate them, which leaves them to doubt other evidence that many languages may be related to one another. Ruhlen is a lumper, as was Greenberg (now deceased), under whom he studied at Stanford University. As we saw earlier, Greenberg classed hundreds of Native American languages into only three groups, whereas most linguists see more than two hundred separate categories for these same languages. However, Greenberg and Ruhlen themselves rejected the lumper-splitter dichotomy, believing that this distinction makes it seem as though splitting and lumping are "styles" of classification rather than fundamental taxonomic distinctions. They have maintained that neither lumping nor splitting is inherently better than the other but that how languages should be classified is an empirical question.

Both Greenberg and Ruhlen have also considered that language families may themselves be related, at a deeper time depth than the separation of the individual languages that compose them. On this basis they have suggested "superfamilies" of languages. For example, Greenberg suggested a superfamily category he called Eurasiatic, into which he grouped Indo-European, Altaic, Uralic, Eskimo-Aleut, and Chukchi-Kamchatkan. Aside from this superfamily, Greenberg's own classification of human languages consisted of eleven families; hence, in his scheme, human languages are classified into a total of twelve linguistic units.

Linguists use different methods to study historical relationships between languages. Some linguists compare languages in pairs, deciding if they are related or not. When a number of languages are seen as related, these linguists then prefer to reconstruct the "proto" language, insofar as possible, from which they derived. Proto-languages can then be compared pairwise for possible relationships. Other linguists dispense with reconstructions of proto-languages and use a method called "multilateral analysis," which simultaneously compares a number of words in many languages. It was this latter method that Greenberg used to arrive at his classification of many languages into superfamilies.

Let us assume for the moment that there are deep historical relationships between many languages and even between language families. How far back could this process go? In 1905, Italian linguist Alfredo Trombetti proposed that all human languages could ultimately be traced back to one single ancestral language. This idea was laughed at by other linguists at the time, and today considerable resistance to this notion remains among scholars. Ruhlen told us, "If anyone went into this area of linguistics, they were seen as crackpots."

Most linguists consider that, even if an original human language existed, we would never know, or could never demonstrate the relationships between this original language and current ones. Languages simply change too fast, in this view. Using glottochronology we can trace back language connections only for about 5,000 years, yet it is presumed that humans have been using language for at least 50,000 and more likely 100,000 years.

A few other linguists have pressed on with this idea, and through their work the whole question of an original human language is now being revived. Among these linguists are some Russian scholars and, in America, Greenberg, Ruhlen, and J. D. Bengston. Cavalli strongly supports their ideas. Using multilateral analysis, these scholars tried to determine whether there were some words that did show similarities among several languages of different families and superfamilies. And indeed they found some, in words for very basic and important things—items such as *water* and *lice* (the latter no longer so important everywhere but probably very significant throughout much of human prehistory). Cavalli notes that these kinds of basic words are often those we learn early in life, and this may explain their greater stability over time. Greenberg also found that a common root word—*tik*—exists with variations in many languages to signify "finger" or "one" and related concepts. In Indo-European this root word appears as *dik* or *deik*, which became "digit" in English and "doigt" in French. Interestingly, a semantic link between "finger and "one" may be posited on the basis of the observation that humans the world over use the index finger to indicate the number one.

This very intriguing idea of a single ancestral human language cannot (or at least cannot yet) be supported, or refuted, with genetic data, but the idea fits in with Cavalli's work in a few key ways. For one thing, Cavalli, as noted above, believes that the major human expansions have been triggered by important innovations—for example, agriculture. When thinking of the very first human migration out of Africa, Cavalli agrees with many others that the key innovation may have been the development of modern human language. An improved human communication made exploration of new areas possible and enabled the migrants to adapt to new environments. Cavalli and others suggest that humans had some sort of language before this time; in fact Cavalli considers it possible, based on cranial evidence discussed by Phillip Tobias, that *Homo habilis* used some primitive form of language. We also know that modern chimps, though unable to speak because of their larynx morphology, are able to learn some sign language. But modern languages are quite sophisticated and complex (and all modern languages are similar in this complexity). Second, Cavalli notes that shortly after the time he presumes modern language to have developed, we also see in the archaeological record a distinctive diversification of tool technology, which indirectly supports the existence of more sophisticated linguistic communication, as has been proposed by G. Issac.

And finally, the idea of a single ancestral human language, gradually diversifying over time as humans carried their language across the globe, is highly consistent with Cavalli's ideas on cultural transmission (as covered in chapter 4). We explore this idea further in the next section. In chapter 6 we also show what new light genetic data on the Y chromosome throws on the issue of an ancestral human language.

Genetic and Linguistic Coevolution

In his book *On the Origin of Species*, Charles Darwin offered an insight about human biological and linguistic evolution:

> If we possessed a perfect pedigree of mankind, a genealogical arrangement of the races of man would afford the best classification of the various languages now spoken throughout the world; and if all extinct languages, and all intermediate and slowly changing dialects, were to be included, such an arrangement would be the only possible one. (Quoted in Cavalli-Sforza 2000:167)

Cavalli claims he had forgotten this passage from Darwin and that it was pointed out much later to him by a colleague, yet his work has gone far to demonstrate that Darwin's insight may have been correct. Cavalli and collaborators Alberto Piazza, Paolo Menozzi, and Joanna Mountain showed in 1988 that by providing a genetic "pedigree of mankind" and matching it up with a linguistic classification of known languages, we see a very strong correlation. Figure 5.3 shows this alignment of human populations used in Cavalli's research, their genetic connections on the left and the linguistic affiliations on the right, using Merritt Ruhlen's classifications. This is the gene / language correlation that so excited Cavalli that he spontaneously drew it out for Ruhlen in the grocery store, as mentioned at the very beginning of this book.

As the figure shows, genetic populations and language families line up well; in addition, populations next to one another in the genetic tree usually speak languages of the same family. Note that in figure 5.3 only sixteen of Ruhlen's seventeen language families are shown. This is because Cavalli's team had no genetic data from the Caucasus populations when drawing this tree. Also shown in this figure is Greenberg's Eurasiatic superfamily. Note that the genetic tree lends support to this language grouping.[3]

---

[3] Not shown in figure 5.3 is another language superfamily called Nostratic that some Russian linguists have proposed. This superfamily covers most of the same languages as Greenberg's Eurasiatic superfamily but in addition includes Dravidian, Afroasiatic, and some Caucasian languages.

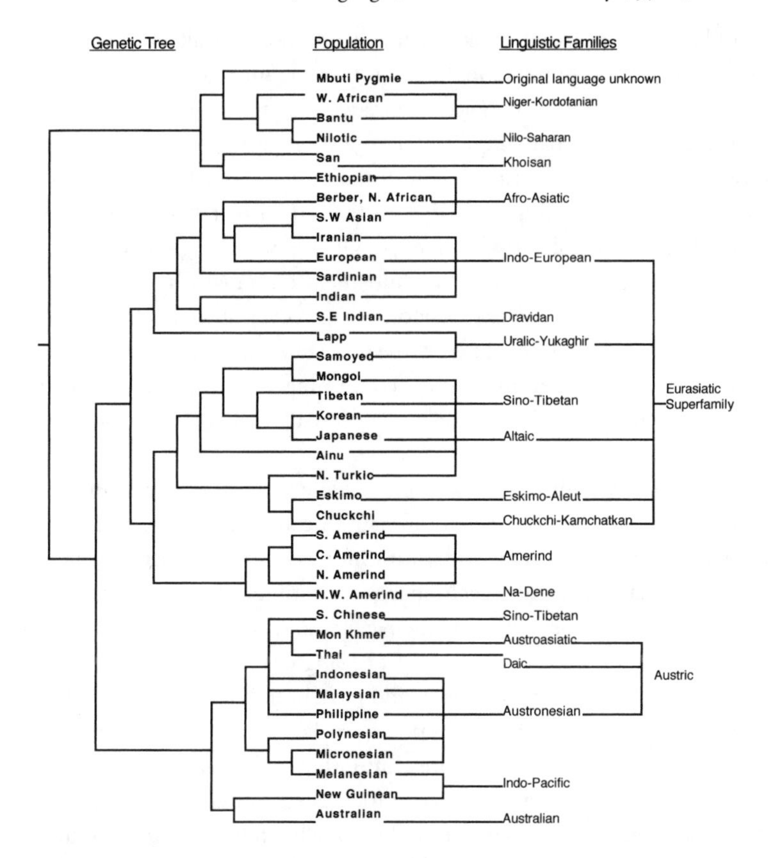

Figure 5.3 Correlation between linguistic families and human populations both shown as phylogenetic trees. (Redrawn from L. L. Cavalli-Sforza. 2000. *Genes, Peoples, and Languages*)

It appears from Cavalli's work that human beings embarked on a number of expansions, taking both their genes and their languages with them, transmitting both over the generations, such that genes and languages each underwent "descent with modification" (to borrow a phrase from Darwin) in a rather parallel way. More on this parallel in a moment, but first we will note some negative reactions to Cavalli's findings.

Cavalli's demonstration of close correlations between languages and genes was strongly criticized by Richard O'Grady, Richard Bateman, and others in terms of data, methods, and interpretations. O'Grady et al. (1989), for instance, had objected that Cavalli's gene frequencies were based on "very small samples, sometimes of single individuals." To this Cavalli and collaborators in a 1990 paper responded, "Presumably they are unfamiliar with the data, methods, and models of human population genetics" (16). In addition, Cavalli and coworkers had used close to 1,800 populations with a total of about 80,000 gene frequencies, a far cry from a "small sample." O'Grady et al. also objected to Cavalli's use of the Greenberg / Ruhlen linguistic classification, especially in regard to the superfamilies, which, these critics point out, are highly controversial and not accepted by a majority of linguists. They further criticized the arbitrary nature of Cavalli's human "populations," while Cavalli and his colleagues pointed out that the nature and distribution of the human species "make it impossible to establish nonarbitrary boundaries" (1990:17). Moreover, in 1992, when statistical methods for correlation between trees became available, Cavalli's team subjected their data to additional tests, confirming the high strength of the correlations. Later, a New Zealand team confirmed Cavalli's conclusions using an independently invented statistical method.

In Cavalli's view, both genetic and linguistic evolution operate through a succession of fissions. People branch off from one another, taking their genes and their languages with them, and with diminished contact between them, their genes and languages will each change over time. The greater the time since separation, the more different their genes and their languages will become (see fig. 1.5). Of course, a human gene / language correlation should not be expected to be perfect. Humans are, and have been, quite capable of skewing such a correlation through various processes. For example, a small group of powerful invaders may overtake an area, imposing its language on conquered peoples. In this case there may be very little genetic admixture between the two populations or perhaps there was only a small demographic contribution from the conquerors, but a complete language replacement of the conquered group may take place within a few generations. A major example of this comes from the European colonization of the Americas, a process that brought about the near replacement of Native American languages with the languages of the conquerors—English, French, Spanish, and Portuguese. Another example comes from the Hungarian language spoken in Central Europe. Hungarian belongs to the Uralic language family. It was brought to Central Europe at the end of the ninth century by the nomadic Magyars from Central Asia, who imposed this language on Indo-European speakers in what is now Hungary. There was little genetic admixture between these two populations, but a total language replacement resulted, so that Hungarians today are genetically largely European but speak a Uralic language. There are many other examples where known historical events can account for a lack of fit between genes and language distributions. A reverse kind of case comes from the Basques, who are genetically quite European, reflecting admixture over thousands of years, but have retained their distinctive language. Basque is thought to be a very ancient language, descended perhaps from the language of the original Cro-

Magnon people who entered Europe around 40,000 years ago. Basques were able to keep their language because of their still visible, strong social cohesion and location in a relatively isolated corner of Europe. All other Europeans repeatedly lost their original language and acquired that of the newcomers. In a 1994 review article Renfrew expanded upon many other mechanisms of linguistic spread and change that will affect the extent to which genes and languages correlate.

The parallel evolution of genes and languages is explained by Cavalli in terms of his theory of culture transmission (see chapter 4). Language is a part of culture that is largely vertically transmitted. We learn language very early in life, mostly from our parents (we call it the "mother tongue," which says a lot about cultural transmission of language). Indeed, after adolescence, learning language is considerably slower and less efficient, so that one cannot easily learn a second language after this stage of life. As a result, language tends to change very slowly compared with other aspects of culture, in particular those aspects that are transmitted horizontally. Like genetic transmission (also vertical), language transmission tends to be conservative. Moreover, the need for humans to understand each other keeps languages from evolving too fast. The reason, then, for the correlation between genetic and linguistic evolution is that both rest on a common cause: vertical transmission (genetic for genes, cultural for languages). This brings Cavalli's work full circle, integrating his genetic data with linguistic data inside his theoretical framework of cultural evolution.

Genes and languages may tend to evolve in parallel, but there are important differences between genetic and linguistic evolution. One is that linguistic evolution (itself slow compared to other types of culture change) is rapid compared to genetic evolution. In another way, according to Cavalli, linguistic and genetic evolution are different, as least as far as the evolutionary origin of languages is concerned. Natural selection, which drives our biological evolution, operates at the level of the individual. Thus, advantageous mutations are differentially favored since their individual carriers are likely to leave more progeny than those who lack the mutation. As for human language, we can see, at least in hindsight, how its development was advantageous, but natural selection could not have favored it at the individual level; at the least, natural selection must have operated on pairs of persons, perhaps a parent and a child, or two or more siblings. Therefore, improvement in language skill may have been selected first in the family, where two or more members inherited the same mutation that improved their language capacity. A single individual with advanced language capabilities, then, is not favored unless he/she can communicate with others. This appears to be, if not quite a "group selection" effect, at least a kind of socio-natural selection operating in human cultural evolution.

In September 2002, Cavalli delivered an invited address to the International Conference on the Origin and Evolution of Languages, held at the Collège de France in Paris. In this address, Cavalli concluded from his decades of work on genes and languages:

> [We] are reaching a consensus on the last 100,000 years of human evolution. It was dominated by the expansion of a small population, probably a single tribe, in East Africa. It probably spoke a single language, as sophisticated as all surviving languages. This gave to modern humans a major advantage in providing communication of better quality than languages available before. Together with other advances it promoted demographic growth and migration which became gradually faster. The expansion of humans to the whole inhabitable Earth favored a differentiation of genes and languages and the two evolutions went on in parallel.

In the next chapter, we temporarily leave the field of anthropology and explore Cavalli and coworkers' breakthrough work on human origins as revealed by DNA markers, which had not hitherto been used in the study of human phylogenies.

Chapter 6

On to DNA Polymorphisms and the Y Chromosome (1984–)

Contributing to bacterial genetics, then mathematical population genetics, linguistics, cultural anthropology, and archaeology would categorize anyone as a modern-day Pico della Mirandola (1463–1494), the famed early Italian Renaissance scholar who was reputed to have known everything that was known in his times. One could argue that, like Pico, Cavalli became a modern Renaissance man (if he was not one before) when he adopted the newly discovered DNA polymorphism technology. Repeatedly, Cavalli's coworkers told us they found his knack for grasping data and his insights into new avenues of research to be astonishing. Not surprisingly the use of DNA instead of proteins as markers of human evolution drew Cavalli's attention as soon as techniques to study DNA polymorphism were developed.

Here we need to review the kinds of polymorphisms, now called "classical polymorphisms," that Cavalli had studied up to this point. Polymorphism is defined as a variation on one "theme." Until the 1980s, the only available genetic "theme" to study polymorphism was proteins. Different organisms in a population make proteins that, although they carry out the same function, can differ slightly in their chemical and physical properties. This is because these polymorphic proteins are composed of slightly different amino acids. All natural proteins are composed of twenty different amino acids (plus two variants) and, ultimately, which amino acids appear in a protein molecule, and in what place, is determined by the DNA gene that codes for that protein. What is more, this gene is composed of a string of base pairs in a double helical configuration, with the base pairs arranged in a very precise sequence. We saw in chapter 1 that mutations alter the nature of one or more base pairs in the DNA, thereby altering the base-pair sequence of a gene. Therefore, mutations also change the sequence of amino acids in proteins. Classical polymorphisms are just that: variations in the amino acid content of a given protein that cause measurable differences in the physico-chemical properties of that protein. These differences can be detected by gel electrophoresis, a technique by which the movements of proteins in an electric field are measured. Polymorphic proteins move at different speeds, an easily detectable property. In the 1960s and 1970s, analytical work on proteins had become routine, and many classical polymorphisms had been discovered. However, one could not study polymorphisms at the DNA level because good techniques to determine the base sequences of genes had not yet been invented.

Why should one study DNA polymorphisms in addition to protein polymorphisms? The answer is simple: over 95 percent of human DNA does not code for proteins. Therefore, studying protein polymorphisms uncovers less than 5 percent of the potential variation that exists in our genetic material. In other words, studying human polymorphism at the *protein* level reveals only a small fraction of the total diversity that exists in individual human genomes. Today, "classical markers" have all but been abandoned in favor of DNA markers. As we saw in chapter 5, studies of polymorphism at the DNA level can be done with RFLP (restriction fragment length polymorphism) analysis, which also involves electrophoresis. Also, we will see that sequencing the base pairs present in DNA provides the highest possible resolution of DNA polymorphisms.

In this chapter, we will see how Cavalli took advantage of the new DNA technology to refine his models of human evolution and migrations over time. Two major themes in his work can be distinguished: the study of DNA **microsatellites** and the study of the Y chromosome. This marks an important period in Cavalli's research; in fact, Cavalli calls the ten years he spent studying the Y chromosome and microsatellites "one of the most important periods of my life." Before covering the specifics of microsatellites and the Y chromosome, we explain what these two elements are.

Microsatellites

Single nucleotide polymorphisms (SNPs; see chapter 5) are just one class of variation among individuals that occur at the level of DNA sequences. Another type of polymorphism is represented by DNA microsatellites. By definition a microsatellite is a DNA sequence composed of blocks of two to six bases repeated many times, as in CACACACACA or ATGCATGCATGC. The length of these repeats is highly polymorphic and varies among individuals. We saw in chapter 5 that scientists use restriction enzymes to cut DNA and measure fragment size. In this case also, researchers can use restriction enzymes to cut microsatellite DNA and measure its length by electrophoresis. Other techniques can be used to estimate the length of microsatellites, but they are not relevant in this context because they yield the same type of information as provided by restriction enzymes. More than 8,000 different microsatellites have been identified in the human genome. In addition to being used in the study of human evolution, SNPs and microsatellites are used in forensics and paternity cases.

Cavalli made abundant use of microsatellites and SNPs in his research on the human family tree. But rather than continuing to study a battery of genes dispersed throughout the human genome, he started in 1994 to focus on a single chromosome—the Y chromosome.

The Y Chromosome

Chromosomes were first observed under the microscope in dividing cells in the 1880s. By using appropriate dyes, biologists observed elongated bodies that seemed to divide and segregate equally into two daughter cells that resulted from the division of a mother cell. A couple of decades later, well after Mendel had put forth his theory of biological inheritance, geneticists reasoned that chromosomes would be good candidates for the physical basis of what were still mysterious genes. After all, the genetic properties of daughter cells were identical to those of the mother cell, and chromosomes were seen to reproduce inside the mother cell before separating, in equal numbers, into the daughter cells. In other words, chromosomes were constant in shape and numbers in all the descendants of a cell. This constancy could potentially explain why genetic traits are constant too.

This hypothesis proved to be correct. By the mid-1920s, geneticists had obtained incontrovertible evidence that genes are located on chromosomes. Then it was shown that chromosomes are rich in DNA, that DNA is the genetic material, and that DNA has a double helical structure. In higher organisms, chromosomes are composed of much more than DNA, however. They also contain vast amounts of proteins (about 50 percent) that provide chromosomal structure and assist in the regulation of gene activity. Chromosome structure and regulation of gene activity are still incompletely understood and constitute important topics in modern molecular genetics. In addition, it was discovered that mammalian male and female cells, including human cells, did not possess identical sets of chromosomes: male cells contain a small chromosome called the Y that female cells lack.

Human cells are equipped with two sets of chromosomes, one coming from the mother and one coming from the father, for a grand total of 46 chromosomes (23 pairs). Exceptions are gametic cells—sperm and egg—which, through the special cell division process that creates them (called meiosis), only carry one set of 23 chromosomes (one per pair). Chromosomes can be subdivided into two categories: the *autosomes* and the *sex chromosomes*. Individuals of both sexes carry

44 autosomes, which are not sex-specific. However, in addition, females have a pair of X chromosomes (they are XX) whereas males have a single copy of the X chromosome and one copy of a very small chromosome called the Y. Thus, mothers contribute one X chromosome to their sons *and* daughters whereas fathers contribute their own X to their daughters *or* their own Y to their sons.

In humans, the sex by default is female. We know this because, in rare cases, some individuals are born with only a single copy of the X chromosome in all their cells (they are called Xo). These individuals are phenotypically female, meaning that in the absence of a Y chromosome, but with only one X chromosome, the sex of a human being is female. It follows that a main function of the Y chromosome is to determine maleness in humans. It is known that the Y chromosome carries the *SRY* gene, whose function is to differentiate into testes the embryonic tissues destined to become gonads in the newborn. Several other genes are also known to be expressed in testes only. Once testes are functional, they start producing high levels of testosterone, the hormone that gives males their other phenotypic characteristics. Without an *SRY* gene, the embryonic gonads differentiate into ovaries. This is known because some very rare phenotypic females are in fact XY, not XX. In these cases, there was an anomalous removal, upon fertilization of an egg by a sperm cell, of the part of the Y chromosome that carries *SRY*. Without this gene, but with the rest of the Y chromosome present, these individuals are born and remain female.

What then is Y chromosome DNA other than just *SRY* and other male-specific genes ? Peter Oefner, one of Cavalli's present collaborators whom we will meet again later, calls the Y chromosome "the junkyard of the X chromosome." And indeed, the Y chromosome contains genes also found on the X chromosome. This portion of the Y chromosome is known as its recombining portion. Since here the Y chromosome contains genes also present on the X chromosome, shuffling of these genes between the X and Y chromosomes takes place through crossing over and recombination. This complicates the analysis of this part of the Y chromosome DNA. Thus, the recombining region of the Y chromosome is less interesting from the viewpoint of population genetics.

However, the vast majority of the Y chromosome consists of long stretches of DNA, most of which contain no genes, and so here the Y chromosome does not recombine with the X chromosome. In that sense, this region remains "pristine." This region, called the nonrecombining region (or NRY), represents 95 percent of the total length of Y. Here the Y chromosome cannot recombine with the X chromosome because it does not share any similar DNA with X. This is the portion of the Y chromosome upon which population geneticists focus their attention.

Also, the Y chromosome has an extremely important property: it is transmitted paternally from father to sons, never to daughters. In a sense, the Y chromosome is biologically a "mirror image" of mitochondrial DNA, which is transmitted maternally (to both daughters and sons, however). Of course, mitochondrial DNA and the Y chromosome have completely different genetic functions and sequences.

Cavalli Embraces DNA Polymorphisms

We saw in chapter 3 that Cavalli had developed mathematical and statistical methods to tackle polymorphisms. Mathematics deals with abstract concepts, and whether it is applied to classical markers (proteins) or DNA markers makes no difference at all except, of course, that DNA permits much finer analysis of variation than do proteins. Therefore, Cavalli's next effort was to refine his models of human origins and migrations, using DNA, but not necessarily leaving proteins behind.

But times had changed. Molecular genetics techniques had become very sophisticated and one could not learn them overnight. Again, it was Cavalli's knack for devising interesting research problems, as well as his charisma, that allowed him to surround himself with bright junior people who were proficient in the DNA analysis techniques. This, by the way, is how cutting-edge science works. A scientific leader does not master every new technical minutia. Rather, a project leader provides historical expertise, plus a great deal of experience, to a scientific topic. He or she then leads a cohort of technically savvy junior investigators in what hopefully is the right direction.

We saw in chapter 5 that, already in the early 1980s, Cavalli had understood the importance of mitochondrial DNA in the study of human evolution. However, as we also saw, he had been "scooped" by Allan Wilson (now deceased) and his team at Berkeley. Wilson was able to make more rapid progress because he had been using a more sophisticated DNA analysis technique than that used by Cavalli's group. But mitochondrial genes represent only a small part of human DNA, only 1/200,000th of it, actually. In addition to mitochondrial DNA, there was still a vast treasure trove of autosomal and sex chromosome–linked genes that remained to be studied.

Cavalli's first article on DNA polymorphism as determined by RFLP analysis was published in 1991 (Bowcock et al. 1991). In it, he studied one hundred single nucleotide polymorphisms in the DNA of five populations from four continents: Europe, Asia (China), Africa (Pygmies), and Oceania (Melanesia). These studies largely confirmed Cavalli's previous model of human origins based on "classical" markers. Interestingly, these results also strongly suggested that Europeans are the end product of an early admixture that took place over 30,000 years ago, between Africans (35 percent) and Asians (65 percent) (fig. 6.1). Although it now turns out that Europe received a greater contribution from Asia, this genetic finding—

which shows the biological sharing, the common ancestry, of several different populations—is a beautiful antiracist statement. These results were obtained in collaboration with Anne Bowcock, formerly a postdoctoral associate in Cavalli's laboratory and now a professor in St. Louis, Missouri. Also among Cavalli's collaborators on this remarkable study were Judith and Kenneth Kidd, now at Yale University, but longtime associates of Cavalli, even dating back to his days in Italy. Later, Cavalli and collaborators drew a phylogenetic tree based on DNA isolated from the cells of over one hundred individuals, using thirty microsatellite markers. The tree separated the individuals according to their continent of origin with a probability of 90 percent, showing that provided enough markers are used, it is possible to trace the geographic origin of single individuals.

There then followed a series of articles exploiting RFLP technology even further. Using microsatellite data, Cavalli's team showed that microsatellites diverge most in Africa, an idea consistent with an African origin for modern humans. Remember that the more ancient the ancestors of a population are,

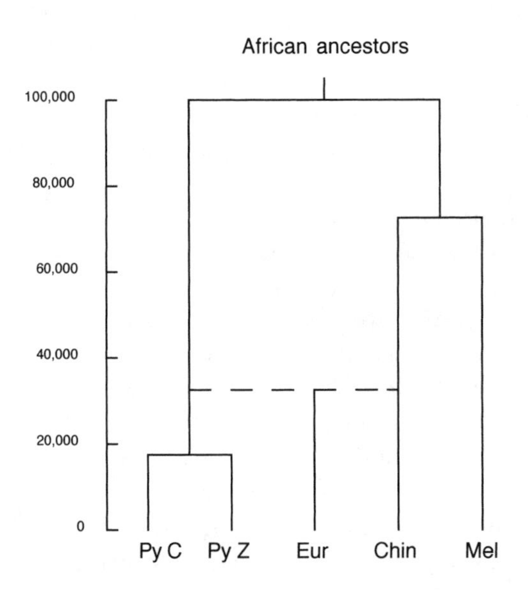

Figure 6.1 Phylogenetic tree showing Europeans as an admixture of Africans and Asians. According to this tree, non-Africans split from Africans about 100,000 years ago. The next split produced Asians (Chin = Chinese) and Melanesians (Mel) about 68,000 years ago. Then, about 30,000 years ago, an admixture of Africans and Asians produced the Europeans. Pygmies (C = Central African Republic; Z = Zaire, now the Democratic Republic of Congo) split from the African branch about 18,000 years ago. (Redrawn from A. M. Bowcock et al. 1991:840)

the more their DNA has had time to change. Further data allowed the team to date the split between Africans populations better than did the mitochondrial DNA data obtained by others. The first split occurred about 100,000 years ago and peopled Africa. Then, the rest of the world began being settled after 40,000 to 50,000 years ago. The "Out of Africa" model for human origins, followed by population splitting, was gaining acceptance.

But then came some criticism. Some scientists remained convinced that mitochondrial DNA was a better key to understanding human prehistory than autosomal markers, be they "classical" or DNA. There may have been several reasons for this. First, one often tends to think that one's research is better than that of one's colleagues. This is not very scientific, but it is a fact of human nature. Second, it is likely that many researchers were unable to follow Cavalli's (and Feldman's) sophisticated statistical analyses. Finally, chromosomes recombine. During meiosis (the cell division that gives rise to gametes), paternal and maternal chromosomes line up very closely. This allows chromosomes to break at identical points and rejoin, or recombine, in order to restore full chromosome integrity. This crossing over causes a problem: gene variants that were originally on a maternal chromosome now find themselves on a paternal chromosome, and vice versa. This problem is tractable, but some scientists felt that chromosome recombination muddied the picture. In addition, at about the same time, some researchers stated that mitochondrial DNA could also recombine. This would also complicate analysis, in particular in rare cases where the male gamete did contribute some mitochondrial DNA to the fertilized egg, or simply because of mitochondrial DNA's heterogeneity. One solution to this problem was to redo the whole research with the nonrecombining portion of the Y chromosome.

The Y Chromosome Breakthrough

As explained, most of the Y chromosome DNA does not engage in gene recombination. This means that, barring rare mutations, this chromosome is "fixed" in time.[1] Therefore, populations that have acquired a particular base sequence at some point in the past will potentially retain that sequence for a

---

[1]. According to the Torah (in Hebrew) or the Pentateuch (in Greek), the first five chapters of what the Christians call the Old Testament, Moses led the Israelites out of Egypt about 3,000 years ago. Soon thereafter, a caste of priests, the Cohens, became the standard-bearers of the new Jewish monotheistic faith. Priesthood was paternally transmitted. It is amazing to realize that, after all these thousands of years, 50 percent of the men named Cohen carry the same gene variants on their Y chromosome (Thomas et al. 2000). Gene dilution via outbreeding—that is, intervention of non-Cohen men (illegitimate paternity)—was minimum in this case.

long time. In other words, a particular base sequence associated with a particular population will be mostly characteristic of that population only. What is more, if that same sequence is found in various geographic areas, it is very likely that direct descendants of the original population that carried that base sequence moved to different locales. In fact, an added advantage of Y chromosome studies is that one does not have to follow the fate of whole populations. Because the Y chromosome is inherited from only one parent (the father), one can build a phylogenetic tree (a pedigree) of single individuals and go back to a single, last common ancestor.

Of course, mutations in the Y chromosome do occur, although rarely, and without them we would not be able to trace the diversification of Y-DNA sequences. For example, if two populations share the same sequence, except at two points, one can assume that the population with two mutations arose from the first one but did not precede it. As already stated in chapter 1, mutations—that is to say, base-pair changes in the DNA—should not be seen as indicative of progress or regress. In fact, most mutations are selectively neutral, meaning that they are neither favorable nor detrimental to the organisms (including humans) harboring them. This is likely the case of mutations taking place in most of the nonrecombining portion of the Y chromosome. Therefore, whether certain human groups show a particular base pair at one spot in the Y chromosome, whereas other populations show a different base pair at that same spot, may mean nothing in terms of fitness of these populations.

These base-pair changes are very rare, and it is therefore laborious to find them. However, this rarity makes the construction of genealogies rigorous and simple, unlike the situation encountered with mitochondrial DNA. One caveat brought to our attention by Walter Bodmer is that mutations in the nonrecombining portion of the Y chromosome may not be entirely neutral. Indeed, this portion of the Y chromosome contains some genes that may have an effect on fitness. As Bodmer puts it: "It may only need one selective marker along there [the nonrecombining part of Y] to pull all the other variations along with it." This "hitch-hiking" phenomenon could then skew the distribution of Y chromosome polymorphisms in different directions.

Normally, determining base-sequence differences among populations means massive sequencing of the DNA isolated from many individuals belonging to these various populations. Isolating DNA from many blood or cell samples is not difficult or time-consuming because the procedure can be automated. But sequencing hundreds of DNA samples in the hope of finding minute base-sequence changes is pure drudgery. As we saw, base changes in the Y chromosome are extremely rare. Some nonrecombining Y polymorphisms had been described before, but they were not easily mapped and were not sharply reproducible. Before Cavalli began his work on the Y chromosome, one excellent marker, called YAP (for *Y* chromosome *A*lu *p*olymorphism), had been discovered by Michael Hammer. The term *Alu* refers to a specific DNA segment inserted at a given location on the Y chromosome. The origin of this insertion was originally, and mistakenly, placed in Japan. Further research based on more data proved that, in fact, this insertion took place in Africa. Cavalli's goal was thus to discover many more polymorphisms to obtain a more precise Y chromosome phylogeny. Therefore it became incumbent upon some of Cavalli's coworkers to do brute-force sequencing with little expectation of quickly finding many sequence changes.

One of these coworkers was Peter Underhill, now a key figure in Y chromosome research and lead investigator, who currently runs Cavalli's lab. Before working with Cavalli, Underhill had applied his skills with automated DNA sequencing at the core lab facility of Stanford Medical School. He did not particularly enjoy his routine job there and wanted more direct involvement in research. In 1991 he was told that Cavalli needed someone with just his qualifications in his lab. After a visit to Cavalli's office and an interview, Underhill was hired and started working right away on mitochondrial DNA and microsatellite sequencing. Underhill knew no population genetics but, said Cavalli, that did not matter; he would learn on the job.

Soon after, Cavalli's interest in mitochondrial DNA began to decrease, largely because it is too variable. Some new research tool was needed, and that tool was to be the Y chromosome. At the time (in 1994), DNA-sequencing machines were much less efficient than they are today, and only a short stretch of 14,000 base pairs of the nonrecombining portion of Y had been sequenced by David Page of the Massachusetts Institute of Technology and his postdoctoral assistant, Douglas Vollrath, now a professor at Stanford. Cavalli thus decided to start more sequencing, using DNA from twenty individuals, four per continent. Cavalli's technician, Joan Hebert, who had already done much mitochondrial DNA sequencing for him, was put in charge of the project under Underhill's supervision.

After a whole year at the bench, doing boringly repetitive sequencing experiments, Joan Hebert found a *single* mutation in the portion of the Y chromosome DNA she was analyzing. Interestingly, this change, an A to G transition, was found in 45 percent of the African Y chromosomes studied. Thus, this transition is a good marker for many African populations because it differentiates them from all the others. We know now that this mutation happened during the westward African expansion of modern humans from their place of origin in eastern Africa. Cavalli wanted Joan Hebert to write the article describing this discovery, but she declined. This job was then offered to, and accepted by, Mark Seielstad, a guest Ph.D. student from Harvard University.

But this was just one single Y chromosome polymorphism that Cavalli's lab was able to pinpoint. One cannot perform statistics on just one single event. Some other, faster technique was needed to discover additional polymorphisms. Such a technique was provided by the timely arrival to Stanford of Peter Oefner. Oefner is a medical doctor (a urologist) from Austria. After practicing medicine in Austrian hospitals and in Madras, India, for a while, he decided, like Cavalli himself, that medicine was not his true calling. During his university education, he had acquired skills in analytical chemistry. Before that, in high school, he had been much impressed by the logical beauty of Mendel's laws of genetics. Much like Cavalli had abandoned medicine to study bacterial genetics, Oefner then started devising ways to facilitate DNA sequencing. To do this, he had developed a new matrix allowing the analysis of DNA by high-performance liquid chromatography (HPLC).

HPLC is an analytical technique based on running a liquid chemical sample inside a little column that allows quick separation (in a matter of a few minutes) of the various chemicals present in the sample. Before Oefner, HPLC had been restricted to the analysis of various chemical compounds, including proteins, but never DNA. One day, then, Oefner was presenting a modest poster describing his technique at a scientific meeting. It so happens that David Botstein, one of the "big gun" professors at Stanford, was attending this meeting and noticed Oefner's poster. In a typical rapid-fire American fashion that still baffles the rest of the world, Botstein decided that Oefner was needed at Stanford because his technique was so novel and so full of promise. Oefner accepted the job. After arriving in Stanford in 1993, Oefner verified his technique and discovered that, by varying the temperature at which he ran his DNA samples, he was able to detect DNA polymorphisms in about three minutes. Moreover, he was able to multiplex his columns so he could run up to six DNA samples at one time. Oefner had just discovered that DNA polymorphisms could be detected more than one hundred times faster than by using the old-fashioned sequencing technique. Nonetheless, Oefner and Underhill did not yet know of each other's presence at Stanford.

So it was, by pure coincidence one day, that Peter Underhill, bored with sequencing, took a break from his repetitive work and attended a seminar given by the then unknown Peter Oefner. At this seminar something clicked. Underhill understood immediately that his miserable sequencing days could be over, and that the Oefner technique would revolutionize his work, as long as he could convince Oefner to collaborate with him. This turned out to be not so easy. Oefner was working on other projects and was not sure that some Y chromosome evolutionary problem was worth pursuing. But Underhill was eventually persuasive enough, and their first experiment worked right away. This was February 1995. They had detected a new Y chromosome polymorphism in minutes rather than weeks. (See fig. 6.2.)

Figure 6.2 Peter Underhill (*left*) and Peter Oefner (Stanford, California, 2002). (From the authors' collection)

Their technique relied on running DNA samples on Oefner's columns at a temperature high enough to dissociate DNA molecules that contained a single base difference (as with the A to G transition and all others) relative to a perfectly matched DNA sample. The profiles they obtained from their **DHPLC** columns (where D stands for denaturing) were unequivocal, as mismatched and matched DNA molecules were clearly separated. The stage was thus set to quickly distinguish base variations in the Y chromosome, the whole analysis time involved being minutes rather than days. Very quickly, they were able to detect dozens of Y chromosome variants and apply rigorous statistical analysis to their findings. The Y chromosome had become, almost overnight, as good a tool, if not a better one, than mitochondrial DNA, to trace human origins and diversification.

This story of the success of the Y chromosome research may sound a little serendipitous, but as Louis Pasteur, the great French microbiologist, once stated, "Chance favors the prepared mind." And prepared they were. First, they discovered a C to T transition present in over 90 percent of the men within the native South and Central American populations examined. In North America, this frequency was about 50 percent. Their dating of this polymorphism indicates that the C to T mutation may have occurred about 30,000 years ago, at or before the time those populations destined to become native Americans were on their way out of Siberia to settle the North American continent. This date, however, was determined to be considerably more recent upon their further research.

The story of the DHPLC discovery is also a lesson in the sociology of science. Cavalli explicitly pointed out to us that he was away from his lab, on an extended stay in Pavia, when this great discovery was made. He has said the same thing in different forums. What is one to make of this declaration? First, let us point out that many academics would not have reacted the way Cavalli did in this instance. They would have taken sole credit for the research, ignoring or downplaying the critical role played by their less visible associates. In this respect, Cavalli is a rather unique gentleman-scientist. His associates we talked to affirmed that Cavalli is far from being the scientific autocrat that one might have expected him to be. On the contrary, he has supported and sponsored his junior collaborators, not thwarted them as is all too common in other places.[2] In fact, Underhill told us that when he and Oefner announced the great DHPLC discovery to him, Cavalli gave them all the financial support they needed to buy top-of-the-line HPLC machines while at the same time staying out their way. As Underhill put it: "With this we were like kids in a toy store."

Having honed their DHPLC skills with the two Y chromosome polymorphisms described above, the Cavalli team discovered hundreds more in a variety of human populations. They were getting ready to apply their statistical techniques to data from the Y chromosome. In the years following 1995, other laboratories, including Hammer's, discovered dozens of new mutations in the nonrecombining part of Y. Readers interested in tracking down these mutations and their discoverers should know that mutations found at Stanford are labeled with an M whereas those discovered elsewhere do not start with an M. What follows is a summary of the Cavalli group's conclusions reached with DNA collected from 1,062 men showing a total of 205 new polymorphisms (only 13 other polymorphisms had been determined previously).

First, they observed that, worldwide, there exist only ten groups of Y chromosome variants, or ten major groups of haplotypes (also called haplogroups—that is, ten general patrilines. They calculated that the original *Homo sapiens* Y chromosome appeared between 140,000 to 50,000 years ago, probably in the northern part of Africa's famous Great Rift Valley, a region that includes Ethiopia, the Sudan, and Kenya. This is the region where the earliest modern humans are found (as fossils). Two African populations still harbor this original chromosome: the Khoisan of the Kalahari desert and some East African populations concentrated in Ethiopia and the Sudan. Therefore, the study of the Y chromosome also supports the hypothesis that modern humans first appeared in East Africa. Thus, "Eve" was an African woman and "Adam"

2. At Washington State University, our own campus, one well-known and locally praised faculty member is so paranoid about his research that he forces his graduate students and postdoctoral fellows to sign upon leaving his lab a declaration indicating that they will never engage in research competing with that of their mentor.

was an African man. Of course, like "Eve," "Adam" should not be construed as an isolated man. He undoubtedly was part of a whole group of modern humans. Also, one should not imagine that "Adam" and "Eve" necessarily lived at the same time.

The Y chromosome data also support a human expansion out of Africa. The scenario goes like this: Modern humans originated in East Africa and started migrating inside Africa between 130,000 and 70,000 years ago. During this period, the Y chromosome acquired one known mutation and thus two distinguishable variants of the Y chromosome appeared. Next, during the early glacial period (50,000 to 45,000 years ago), four more Y chromosome variants appeared for a total of six. During that time, humans had migrated to the Middle East, India, and Southeast Asia, and had also reached New Guinea, Australia, southern China, and Japan, traveling via the Horn of Africa and the Levantine corridor. These humans were not necessarily great navigators to have reached Australia. At the time, New Guinea and Australia were part of a single landmass and the expanse of open sea they had to cross was shorter than today.

Then, 45,000 to 30,000 years ago, two more Y chromosome mutations occurred, bringing the total of variants to eight. It is at this time that some populations concentrated in the Middle East dispersed into Europe and Central Asia, taking their new Y chromosomes with them. Other groups went back to Africa. By 30,000 to 20,000 years ago, an additional two Y chromosome mutations took place, to give us the current total number of ten variants. These two mutations took place in Central Asia, and their bearers moved into Europe, in a second wave of migration. They also traveled to Siberia and the tip of Northeast Asia.

By 18,000 years ago, the world had become occupied by humans, with the exception of Polynesia and possibly the New World. It should be noted that all these waves of migration did not necessarily take place in totally unoccupied territory. For example, we saw that Eurasia experienced at least two major similar expansions that took place several thousand years apart. The two last steps of human colonization of the whole planet took place between 30,000 and 12,000 years ago and during the last 6,000 years, respectively. These migrations brought humans from Siberia to the Americas and from New Guinea and Southeast Asia to Polynesia. Figure 6.3 summarizes what Cavalli and his son Francesco have called "the great human diasporas."

Interestingly, Y chromosome data in Europe and the Middle East indicate that about 28,000 years ago, an exponential population expansion must have taken place. This timing corresponds well with the emergence of the Gravettian stone tool technology (28,000 to 22,000 years ago) and the disappearance of the Neanderthals. As we saw, there is presently no evidence that the Neanderthals disappeared because they interbred with modern humans. On

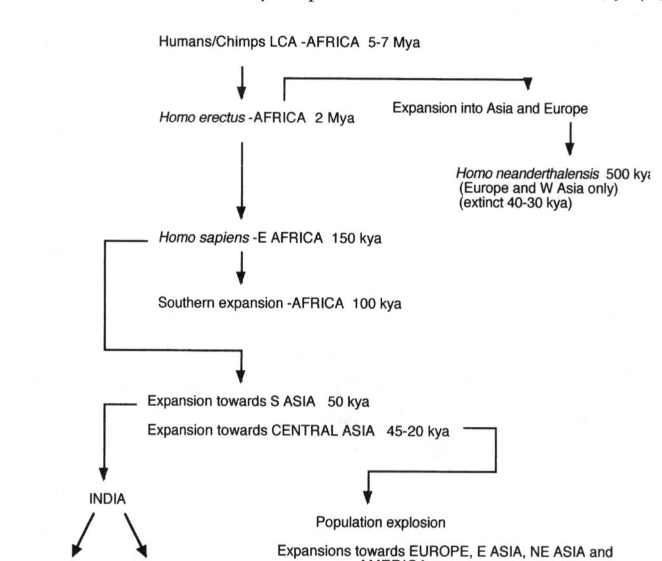

Figure 6.3 Summary of the origin and migrations of the genus *Homo*. LCA: last common ancestor; this LCA gave rise to Australopithecines (not shown), the first hominids, which themselves gave rise to *Homo erectus*. This species was, as far as we know, the first *Homo* population to leave Africa. According to the uniregional hypothesis, descendants of *H. erectus* became extinct everywhere, except in Africa. The link between the LCA and *H. erectus* is entirely based on paleontological evidence, not DNA studies. The spread of *H. sapiens* from Africa throughout the world is based on the work of Cavalli's group and that of others (see text). Mya: millions of years ago; kya: thousands of year ago.

the contrary, studies of mitochondrial DNA (initiated in the laboratory of Svante Pääbo in Germany) from three Neanderthal fossils indicate no presence of their genes in our own mitochondrial DNA. Whether genocide by *Homo sapiens* or changing ecological conditions brought about the demise of *Homo neanderthalensis* is unresolved.

Cavalli and coworkers then discovered that a Y chromosome polymorphism very prevalent in Europe was also found in about 30 percent of male Australian aborigines (the statistical uncertainty is due to small sampling). This meant that Australian aborigines and Europeans are much more closely related than initially suspected. However, the explanation is historical, not prehistorical. In the eighteenth and nineteenth centuries, Great Britain got into the habit of banishing its convicts (mostly men, often jailed for debt) to their new, very distant, and largely empty Australian colony. The rest of the story is easy to imagine: local women clearly interbred with these convicts.

Another benefit of the Y chromosome research concerns the relative degree of migration done by men and women. One might think that both genders moved about equally in the past. However, this view is not supported by Cavalli's results. When the geographical concentrations of Y chromosome, mitochondrial DNA, and autosome polymorphisms were compared, Cavalli and coworkers discovered that Y polymorphisms are more localized geographically than mitochondrial DNA and autosomal polymorphisms. This is consistent with the view that males migrate, genetically speaking (i.e., leave the social group or their residence), less than females. Interestingly, research with chimpanzees shows the same trend: it is the females, not the males, that tend to leave the group most frequently. By doing so, the females spread their mitochondrial DNA and their autosomes widely, but the male Y chromosome remains geographically localized, maintained in place by stationary males. In humans, it is known that 70 percent of all cultures practice patrilocality—that is, the grooms usually stay put while the brides move to join their husband's household. This tradition may be very ancient, but there exist numerous exceptions. For example, in some cases immigration to new regions is done mainly by males, who take local wives, as happened relatively frequently during the recent expansions of Europeans to the Americas and Australia.

Finally, statistical analysis of the Y chromosome genealogy was also used to gain an estimate of the number of humans that existed about 100,000 years ago, and from whom present human populations are all descended. The Stanford group showed that this number is about 1,000. Thus, we all are the descendants of an African tribe, composed of roughly 1,000 individuals, who expanded out of Africa and who, over many generations and genetic diversification, colonized the whole world. With this, could it also be true that the languages spoken today on the planet are all derived from the primeval language spoken by this tribe (as discussed in the previous chapter)? Cavalli suggests that, indeed, this might be the case, and his colleague Merritt Ruhlen agrees. When we asked Ruhlen whether we would ever know what language "Adam" and "Eve" spoke, he replied that the Khoisan tongue, with its characteristic clicks, is probably as close as any modern language can be to the primeval language. Anthropologist Joanna Mountain, one of Cavalli's colleagues, even stated that "clicks were once universal."

At this point, one may have noticed that mitochondrial DNA data and Y chromosome data, together with archaeological and other genetic data, present a coherent and consistent picture of human origins. However, mitochondrial DNA data tell us that "Eve's" birthdate was about 143,000 years ago, whereas "Adam's" birthdate, as per his Y chromosome, is put at a most probable date of

93,000 years ago. Furthermore, Cavalli's own results indicate that a split among Africans took place before 100,000 years ago. There is definitely a problem here, as "Adam" and "Eve," while not necessarily being contemporaries, must both have existed in Africa before any population split occurred. How can one reconcile these results? First of all, we need to consider mutation rates and their relatively high statistical error. As explained, knowledge of the mutation rate of DNA is the most commonly used approach to date genetic diversification. The Y chromosome mutates very slowly whereas some parts of mitochondrial DNA mutate rapidly. It may be that better knowledge of mutation rates will bring these two "birthdates" closer together. Second, it may be that a greater bottleneck effect (see chapter 3) at the level of men than at that of women does not allow us to see the real "Adam," whose Y chromosome was eliminated earlier by random drift. If indeed, at one point in our prehistory, there was an acute shortage of men, we may assume that polygyny (the marriage of a man to two or more women at the same time) may have been prevalent about 100,000 years ago, as it still is today in some areas. Certainly, polygyny is much more prevalent today than polyandry (the marriage of one woman to two or more men), which is rare; it is practiced, for example, in parts of Tibet, Nepal, and India. To illustrate the effect of polygyny on genealogies, consider the following example. It has been stated that in a certain village of the Amazonian jungle all the children were fathered by the chief (at least, this was his opinion). If the whole world were reduced to that single village, the chief would be "Adam," and he would be just one generation old. Clearly, more studies are necessary, in particular in-depth comparisons between the two uniparental genetic systems, mitochondrial DNA and the Y chromosome.

Finally, one may wonder whether the "classical" polymorphisms used earlier by Cavalli and coworkers, and their microsatellite and Y chromosome results regarding human variation and migration, are in agreement. The agreement is close, although inevitable refinements were made as the techniques themselves were improved. It is important to emphasize that similar results obtained by three different methods (protein polymorphisms, microsatellites, and the Y chromosome) make an even stronger case for the model proposed. Also, data on the Y chromosome and microsatellite data complement one another. Microsatellites and the Y chromosome do not mutate at the same rate. The Y chromosome mutates slowly and is thus best to determine genetic events that happened in the distant past. On the other hand, microsatellites mutate more rapidly and are best suited to study events that took place in the past 5,000 to 10,000 years. If data from these three sources had produced contradictory results, revisions of the initial hypotheses would have been required (such as, for example, hypothesizing that modern humans did not appear in Africa or did not migrate the way we thought they had, or hypothesizing that modern humans appeared about 6,000 years in the past, as proposed by Bishop Ussher centuries ago). This has not happened, and the model for human origins proposed by Cavalli and his numerous collaborators is presently quite convincing. In fact, Cavalli, imitating subatomic particle physicists, likes to call their interpretation the "standard model" of human evolution,[3] meaning that it seems basically true and open for testing and continuous improvement.

The Y Chromosome in Europe

As we saw in chapter 4, Cavalli's demic diffusion model for the spread of agriculture in Europe was criticized by Bryan Sykes in his book *The Seven Daughters of Eve*. For Sykes and his collaborators, the Neolithic farmers who spread agriculture in Europe, in a wave of advance from somewhere in the Middle East, did not replace the populations of Paleolithic hunter-gatherers

---

3. We have focused in this chapter mostly on Cavalli's recent results and his "standard model" of prehistoric human expansions. A review of some of his previous reports shows that the dates for population splits given in the standard model were initially thought to be different. Based on classical polymorphisms, it was shown in 1988 that the spread of modern humans in Africa occurred about 100,000 years ago. In 1991, based on RFLP analysis, the date of the split between African and non-African was also thought to be 100,000 years ago. In 1995, based on microsatellite analysis, this date became 156,000 years ago (with a confidence interval spanning from 75,000 to 287,000 years ago). Therefore, the literature gives a great variety of dates, and it is legitimate to wonder what to make of this. Estimates unaccompanied by a standard error are not very useful anymore. Confidence intervals like the one given above are usually calculated with a probability of 95 percent. Standard errors are usually half as large and correspond to a lower probability, about 64 percent. At the moment of this writing, the best (and still unpublished) estimates from the Stanford laboratory are 103,000 years (with a standard error of 20 percent) based on Y chromosome data and 168,000 years (with a standard error of 15 percent) based on mitochondrial DNA data.

The difference between these two estimates probably depends on just one parameter, the variation of the number of children per individual. This number is strongly influenced by a common custom in sub-Saharan Africa: polygyny. Although polygyny varied greatly in fourteen countries studied in the 1960s, data showed an average of 134 married women per 100 men. It is reasonable to think that polygyny was more prevalent in the distant past.

It can be expected that more research may be able to refine dates and narrow confidence intervals. Also, it is clear that genetic data cannot be viewed in isolation and should be combined with archaeological and anthropological-demographic observations, as convincingly shown by Cavalli. Also, as pointed to us by Walter Bodmer, the integration of the massive amounts of data on mitochondrial DNA, Y chromosome DNA, and autosomal DNA, together with classical polymorphisms, is just barely beginning.

they met on the way. For them, based on mitochondrial DNA data, extant Europeans have genes that are about 20 percent Neolithic (the genes of the farmers) and 80 percent Paleolithic (the genes of the hunter-gatherers). One must note that, at first, Sykes had concluded that Neolithic farmers had contributed at most 4 percent of their genes, a value that he subsequently corrected. Thus, for Sykes, one should not conceive of the Neolithic farmers as migrating en masse and replacing Paleolithic hunter-gatherers in such a way that little genetic trace of the Paleolithics would be left in modern Europeans. One problem with Sykes's book is that it attributes to Cavalli this idea of Neolithic farmers massively overtaking Paleolithic hunters. But in fact, Cavalli never said or wrote anything quite like that (as we saw in chapter 4). On the contrary, based on principal component analysis, Cavalli and his coworkers concluded that the Neolithic farmers contributed only 26–28 percent to the gene pool of extant Europeans. This result was based on autosomal classical markers, not on DNA. Thus, there is basic agreement between classical markers and mitochondrial DNA markers.

Nevertheless, mitochondrial DNA data, even though they are in rough agreement with Cavalli's classical markers, do not provide as detailed a genetic map of Europe. If principal component analysis is applied to mitochondrial DNA sequences, the map looks very banal and the genetic diversity is low. This is in contrast with Cavalli's results and carbon-14 dating of archaeological sites, both of which provide much greater detail regarding the advance of agriculture and Neolithic farmers over time and space. There may be several reasons for this. First, the portion of mitochondrial DNA (called the D-loop) that Sykes and his collaborators studied has a very high mutation rate. This means that a given haplotype can mutate into another one, and then mutate again to restore the old haplotype. In a situation like this, it is very difficult to build phylogenetic trees, and indeed, Sykes and coworkers eventually admitted this problem. Second, mitochondrial DNA traces the lineage of women, who, we have seen, tend to migrate more than men do. In view of this, it may be the case that mitochondrial DNA and classical marker data were in agreement purely by accident. More research was necessary to confirm or refute the interpretation that Neolithic farmers contributed about 20 percent of their genes to the current European gene pool.

By the time Sykes had come up with his results, Cavalli's lab had developed the DHPLC technique to detect polymorphisms in the Y chromosome. The team was ready to investigate the question of the origins of European males. Cavalli and his collaborators discovered that just ten different mutations on the Y chromosome characterized the vast majority (over 95 percent) of European men.

When these mutations were dated with microsatellites, the ones of mideastern origin were in part older and in part more recent. Using only mutations of most recent origin, in order to get a more conservative estimate of the proportion of mideastern farmers participating in the expansion, Cavalli and collaborators calculated that 22 percent of European farmers were of mideastern origin. This was in reasonable agreement with the 20 percent estimate from mitochondrial DNA and with the 26–28 percent estimate obtained from the study of classical polymorphisms and principal component analysis. Thus, the inventors of agriculture have contributed at least one fifth to one quarter of the genes present in people of European origin. We will see later that this may be somewhat of an underestimate. Interestingly, the Neolithic genes are more prevalent among Mediterranean populations than among other Europeans.

Finally, the rest of the haplotypes showed that at least 78 percent of the European Y chromosomes were derived from much older Y chromosomes that appeared in Europe in two waves. The first one was tentatively dated to about 30,000 years ago and corresponds to a migration of people entering Europe from the East. This Y chromosome carries the ancient Euroasiatic marker described above. Intriguingly, another wave of migration then took place about 22,000 years ago and consisted of people who migrated out of the Middle East. These were still Paleolithic people, not the Neolithic farmers that started migrating out of the Middle East about 10,000 years ago. These Paleolithic populations were forced by the advance of glaciers to find refuge in what is now the Basque country. At the end of the last glaciation, about 13,000 years ago, these people expanded into Central Europe, as hypothesized by another researcher, Antonio Torroni. Later, other populations moved into northern Europe from across the Urals, introducing a language that would later diverge, to become Saami (formerly called Lapp), Finnish, and Estonian.

But this may not be the end of the story. Cavalli told us that the agreement between the mitochondrial DNA data and the Y chromosome data surprised him somewhat because of the unreliable variation of the stretch of mitochondrial DNA studied by Sykes, and the possible differential migration of men and women. It turns out that in August 2002, a British-Italian team under the direction of L. Chikhi revisited one more time the question of the Neolithic demic diffusion model explained in detail by Ammerman and Cavalli in 1984. To do this, they used the Y chromosome results published by Cavalli's team in the year 2000 (Semino et al. 2000) and reanalyzed them with a new statistical technique they developed, called the *Markov chain Monte Carlo technique*.

At the time we interviewed him in August 2002, Cavalli drew our attention to the Chikhi article, which had just been published that month. He was pleased by it because this paper strongly supported his demic diffusion model of agriculture into Europe, but he was also puzzled by it because Chikhi and his coworkers' percentages indicating the genetic contribution of the Neolithic farmers to Europe were significantly higher than what Cavalli himself had proposed. For Chikhi, Middle Eastern Neolithic farmers had contributed as much as about 50–65 percent on average to the current European gene pool, with even higher values among Mediterranean populations, the first ones to be influenced by the Neolithic expansion. These higher numbers may be the result of Chikhi's utilization of Basques and, in particular, Sardinians as references, as descendants of Paleolithic people. Traditionally, Sardinians had been considered "outliers" with regards to their gene frequencies.

About a year later, having thought more about the issue, Cavalli communicated to us the following. First, let us take Chikhi's results at face value: Neolithic males contributed about 50 percent to the gene pool and females contributed about 20 percent. Thus, the Y chromosome bearers (men) contributed more than the mitochondrial DNA transmitters (women). This, says Cavalli, is somewhat analogous to the Bantu/Pygmy situation (see chapter 3), in which there is an advantage for farmers (Bantus) to have more wives than hunter-gatherers (Pygmies). Indeed, much farming work is done by women and, further, hunter-gatherer women are often accepted as wives by farmers. The reverse is exceptionally rare. Also, polygyny is definitely more frequent among Bantus than among the hunter-gatherers. Thus, by analogy, a 50 percent male Neolithic contribution versus a 20 percent Neolithic female contribution (meaning an 80 percent female Paleolithic contribution) is acceptable. This is also the position taken by Walter Bodmer, who sees European migrations as mostly male-dominated.

Moreover, Cavalli points out that Chikhi's results are based on *all* the Middle Eastern haplotypes, including those of less recent origin. However, still unpublished work by Cavalli's group indicates that Turkey, not the whole Middle East, must have been the natural source of farmers who went to Europe. Thus, if it is correct that mitochondrial gene flow was 20 percent and Y chromosome gene flow was 50 percent, it follows that the autosomal value should be the average, or 35 percent. According to Cavalli, it is 26–28 percent. However, Cavalli points out that this is a very conservative estimate, because it is based on the present gene gradient going from southeastern to northwestern Europe. According to him, unless gene gradients are maintained by natural selection, those determined entirely by admixture will always flatten with time. Therefore, Cavalli has always resisted using gene gradients as an estimate of the relative importance of cultural and demic diffusion. As we saw earlier, Cavalli came out with the 26–28 percent value to counteract the original Sykes value of only 4 percent Neolithic contribution.

Further, simulations by L. Sgaramella-Zonta and Cavalli-Sforza (1973) showed that while the rate of archaeological advance is almost perfectly linear, as expected from Fisher's theory of the wave of advance, the expected genetic gradient falls like a negative exponential, and the proportion of admixture due to demic diffusion is expected to be close to zero in northern Europe, as it is found to be. Let us remember that Cavalli's 26–28 percent admixture value was based on the first principal component of gene distribution in Europe, which is essentially a gradient. Since gene gradients are expected to flatten over time, his value should be an underestimate of the true proportion of average admixture between Neolithics and Paleolithics. This explains at least part of the difference between the observed 26–28 percent and the expected 35 percent. However, points out Cavalli, the errors on these numbers are very difficult to estimate and more research is needed.

Cavalli explains that in Europe one notices a rapid advance of the Neolithics from the southern part of Central Europe to the North Sea, probably along the rivers, which always offer rapid transportation. The Neolithic artifacts found along the way are remarkably homogeneous, such as ceramic containers (called Linear Pottery) and houses that are long, with several fireplaces, and might well have housed polygynous families. To continue the analogy with Bantu farmers, the wives of the same man often live in the same house or adjacent houses. Finally, Cavalli states that, "Better data will probably improve these estimates, but it is not unreasonable to think that about one third of European genes derive from farmers of the Fertile Crescent because of demic diffusion." As of September 2004, this is the extent of the discussion on the appearance of farming and new technologies in Europe during the Neolithic period.

In this book we have covered the latest advances in the molecular population studies of human evolution. All the models dealing with prehistoric human migrations rely on lab work, in particular work done with DNA extracted from human beings. But how do geneticists analyze these polymorphisms? How can they be confident about the interpretations of data (at least temporarily, as we have seen repeatedly)? To answer these questions, let us look at a hypothetical scientist whose interest it is to study human origins by using genetic techniques. This is not trivial because, after all, geneticists possess the instruments and the knowledge to tell us who (and even what) we are genetically.

A Day in the Life of a Molecular Population Geneticist

What is it that geneticists do in their labs that allows them to trace the origins and mutations of the Y chromosome, among others? The following is a somewhat idealized description of how things proceed on a typical day.

First thing in the morning, the researcher takes out of her freezer the human blood or cell samples that she is going to process. The blood was most probably harvested by somebody else. The blood or cell samples stored in the freezer can be quite small, because there exists a technique called polymerase chain reaction (PCR) that can amplify very small DNA samples in the test tube. The researcher then uses a commercially available kit containing all the reagents necessary to purify DNA from the blood or cell samples. This step takes a short time. Next comes the PCR amplification step of the gene or genes that the researcher wants to study. Here again, a commercial kit provides all the reagents. The PCR solution containing the DNA is then put inside a thermal cycler, which is a benchtop machine that raises and then lowers the temperature of the samples in a cyclical manner and allows the synthesis of the specific gene(s) of interest. This step takes a few hours and is completely automated. If the researcher has no experiment proceeding in parallel, she can take a break. After the cycling has ended, the amplified DNA samples are purified (this takes minutes) and run through the DHPLC machine.

Things look a little more impressive at this stage. The DHPLC equipment is big enough to cover a dining room table, and is connected to a video monitor that visualizes the DNA patterns as they come out of the column, the latter being hooked up to two big pumps that run the samples through the column. This step takes about three minutes per sample. One of two things can happen next: if the DHPLC pattern shows no polymorphism, this DNA sample is not studied further. If, on the contrary, polymorphism is detected, this sample is interesting and processed further. At Stanford, single nucleotide polymorphisms can be determined in a matter of a few more minutes by running a polymorphism-positive sample through another machine called a mass spectrometer. This truly is a very complicated piece of equipment, able to analyze groups of atoms, depending on their mass and electric charge as they travel through a magnetic field which deflects them. Fortunately, this machine is also completely automated, albeit very expensive (Cavalli's group does not own one, but Peter Oefner does). At the end of the run, the nature of the polymorphism is known, and the real work starts.

Indeed, the analytical portion of the research described above is not exciting. What was engaging at one time was the development of all these sophisticated techniques. But now, the technical aspect of the research is highly simplified by reagent kits and automated equipment. What is now really exciting is putting the data together. Computer algorithms are used to run the statistical and mathematical analyses of the DNA data, evolutionary trees are built, and the results are discussed. Finally, if all goes well, the results are interpreted, critiqued, and eventually written up for publication. In our opinion, this is the most creative moment of the whole process because the researchers' efforts are coming to fruition. As Underhill told us: "You can tell a story with that stuff." It must indeed be enthralling to see a Y chromosome phylogeny emerge from the computer and put on a map of the world. One might even at such a moment spend a few minutes thinking about the great advances made in science that allowed this peek into the distant past. One might also think about the distant ancestors to whom we are all so directly linked.

Remaining Connected with Past Research

While doing all this work with DNA polymorphisms, Cavalli did not abandon the more classical approach to studying human variation that he had started in the 1950s. With Italian colleagues, he continued his in-depth study of human genetic variation in Italy. For this, he used an experimental tool that may come as a surprise: the telephone directory! It turns out that people transmit to their progeny not only genes but also their surname (or last name or family name). Surnames are almost always patrilineally transmitted to children. Since, at least in Italy, women tend to adopt their husband's name upon marriage, a study of surnames over time and space is a study of "genetic" transmission from fathers to sons but not to daughters. In a very real sense, surnames are transmitted like the Y chromosome. The main difference between Y chromosome research and phone book research, however, is that expensive and complicated PCR, DHPLC, and mass spectrometry equipment suddenly become unnecessary, and one does not have to learn how to use these machines! But then, how good is a surname as a "genetic" marker?

As stated before, a study of human evolution is greatly facilitated if the genes used are selectively neutral. In other words, focusing on selectively neutral genes allows us to downplay the role of natural selection, which is often difficult to understand and quantify. Then, only three phenomena affect gene frequencies in populations: mutation, migration, and random drift. Mutations, being rare, can usually be ignored, but illegitimacy is a potential source of concern. But an additional advantage of surnames is that they rarely mutate (except at Ellis Island, in the olden days). Cavalli realized that a study of the distribution of surnames might enable researchers to evaluate the impact of drift and migration on gene frequencies in Italy. For this reasoning to work, one must assume that surnames are selectively neutral or do not confer any advantages or disadvantages regarding survival and offspring production.[4]

---

[4]. Surnames are not necessarily always selectively neutral. Wars, racism, and xenophobia can affect one's fate simply based on what one's name is. A well-known example is anti-Semitism, often based on name alone, not so much on looks. Other examples might include francophobia, anglophobia, americanophobia, etc. wherever they exist. On the other hand, some names can have positive effects, as in Nepal and India where names can reflect membership of a given caste.

With coworker Gianna Zei, Cavalli tabulated over 500,000 consanguineous marriages and over 18 million telephone directory entries to test the genetic drift hypothesis in the Italian island of Sardinia. Parish books and permissions granted for consanguineous marriages allowed them to measure from historical Roman Catholic Church records the variation in time of surname frequencies, whereas telephone directory entries allowed them to measure the dispersion of surnames in space. They then superimposed their statistical results on surnames over several actual gene frequencies they had determined earlier. The match was excellent: both studies agreed that the best way to account for surname dispersal and geographic gene-frequency variation was to conclude that drift played a major role in both. Thus, most genes in the population studied were selectively neutral, just like surnames. There were some exceptions: for example, the gene frequencies for glucose-6-phosphate dehydrogenase (G6PD) did not fit the drift model. This however, was to be expected as this gene is involved in resistance to malaria which, until a few decades ago, still existed in Sardinia. Thus, the G6PD gene is not selectively neutral and one does not expect its frequency to depend on random drift.

A few years later, C. R. Guglielmino joined the group of Cavalli's coworkers to extend this type of study to the other large Italian island, Sicily. There, both surnames and given names (first names or Christian names) were included to address genetic effects as well as cultural practices (first names are inherited only loosely from grandparents, unlike surnames). Cluster and principal component analysis showed that four "genetic" regions could be distinguished in the Sicilian landscape. This agreed very well with the history of the island, and these four regions may correspond to the four principal waves of migration that the island experienced. First came the Elymians whose origins are not known conclusively. Then came the Sicani, who may have been of Lybian or Iberian origin. Third came the Siculi, who originated from the Italian peninsula, and finally, the Greeks in the eighth century B.C. The Arabs, who conquered and held Sicily between the seventh and tenth centuries A.D., contributed less to the island's demography but helped improve agriculture. We can see here again that Cavalli always wanted to put his genetic data (even be they from the phone book) in a historical and cultural context.

But then, drift is only one factor that affects gene frequencies. What about migration? Again, Cavalli and Gianna Zei used church records (540,000 consanguineous marriages) and phone books (10,473,727 entries) covering all of Italy. From this, they calculated male migration rates in all of Italy and provided an accurate measure of drift in 8,000 Italian communes. Of course, modern Italian drift and migration patterns do not necessarily reflect rates and distances of migration of Paleolithic, Neolithic, or even worldwide modern humans. However, research involving birth records, marriage certificates, and telephone directories is orders of magnitude cheaper than analyzing tens of millions of DNA samples. Thus, Cavalli's surname approach is theoretically applicable to the whole world, wherever patrilineal transmission of surnames is the rule, and where telephones exist. Even in places where patrilineal transmission is not practiced—for example, among the Navajo, where children assume the clan name of the mother—similar (matrileanally based) research could be conducted and might yield very interesting results.

In the preceding three chapters and this one, we saw how Cavalli's work in the fields of population genetics, anthropology, and linguistics, conducted over several decades, has presented an exciting and crisp picture of human physical and cultural evolution. As one can well imagine, such an encompassing portrait of the origins and fate of our species might not please everyone. As we saw, cultural anthropologists (but not archaeologists) have almost completely ignored Cavalli's theory of cultural evolution. But that is not all; Cavalli has opponents who are strongly critical of one of his most ambitious projects, the Human Genome Diversity Project (HGDP). As usual in science, more data are necessary to refine models and explanations that scientists build. In that vein, Cavalli has been striving for several years to gather more information regarding genetic diversity in small, isolated indigenous populations worldwide. This effort is at the core of the HGDP. We will see in the next chapter that, contrary to Cavalli's previous scientific endeavors, the HGDP has met with significant societal opposition.

Chapter 7
The Human Genome Diversity Project (1991–)

In one year, Cavalli's career both loomed toward new promise and fell into turmoil. This was 1991–92, one of the most turbulent periods in Cavalli's life. He had begun to organize his most ambitious project—the Human Genome Diversity Project (HGDP), a program to collect and analyze DNA samples globally. This was to produce a mine of data to comprehensively explore human prehistory, determine the genetic relationships between the earth's populations, and provide valuable information on human genetic diseases. For this project Cavalli teamed up with Allan Wilson at Berkeley (of "African Eve" fame; see chapter 5). In 1991 these two—along with Mary Claire King, who took the initiative—and others put out a call for a worldwide survey of human genetic diversity, including samples from many indigenous populations. They received initial enthusiastic responses from anthropologists and potential funding agencies. Then Allan Wilson died of leukemia, and later in the same year Cavalli suffered his heart attack and underwent by-pass surgery. While still in recuperation from this, all hell broke loose: in his efforts to launch the HGDP, Cavalli was accused by some groups of being a racist vampire!

Before showing how and why this happened, we will look briefly at the history of human DNA sequencing and note what the DNA sequencers have achieved so far. In the following we will also make clear the distinction between the Human Genome Organization (HUGO), the Human Genome Institute (a National Institutes of Health—NIH—institute), and Cavalli's brainchild, the HGDP.

The Human Genome Projects

One of the grandest scientific projects of the twentieth century, founded and initially directed by James Watson, is the full understanding of the functioning of the some 25,000 genes that human beings carry. A rough draft of the human genome was obtained in parallel, in the year 2000, by a private company (Celera) under the direction of Craig Venter and a competing multinational association of laboratories under the direction of Francis Collins in the United States and the Wellcome Trust–supported Sanger Institute in the United Kingdom, just to name the two institutions that contributed most of the data. To celebrate the occasion, President Bill Clinton invited the two directors to the White House for an official announcement in front of television cameras. Today, even though practically all of the 3.1 billion base pairs of human DNA have been sequenced, we are still far from understanding what all this DNA does. Of the expected 25,000 human genes, only a small fraction of their function is known. Thus, the work of the human DNA sequencers is not finished, and it will take years to annotate—to position and attribute a function to—all our genes.

The concept of sequencing the human genome goes back to the 1980s. Indeed, in those early days of DNA sequencing, several scientists were already thinking about deciphering the human genome. It took almost a decade to establish strategies and secure the funding that culminated in the sequencing of human DNA. The concerted effort to sequence the human genome was begun in 1988 and, in the beginning, was mostly privately funded, in the United States by the Howard Hughes Medical Institute and in Great Britain by the Imperial Cancer Research Fund. Soon thereafter, public funding (from the NIH and the Department of Energy, and coordinated by the Human Genome Institute) became available to laboratories doing human DNA sequencing in the States as well as Europe. As mentioned earlier, the Human Genome Organization (HUGO), with Walter Bodmer as its second president, was also active as an association of scientists interested in the deciphering of human DNA. It has a small office in London, and its main purpose is to organize yearly symposia where results on the human genome are discussed. Twelve years after the initiation of the project, the human genome had been sequenced.

Meanwhile, in 1991, Walter Bodmer had asked Cavalli to chair a committee, which Cavalli called the Human Genome Diversity Project, or HGDP. Because of his faltering health at the time, Cavalli was helped in the beginning by Kenneth Kidd, Kenneth Weiss, Marcus Feldman, and Mary Claire King. This group received funding from the NIH's Institute of General Medical Sciences, the National Science Foundation, and, at first, the Department of Energy, to organize several symposia to discuss the logistics of studying the DNA of diverse human populations. Four such symposia were held. The first one convened at Stanford University and was attended by population geneticists; James Crow chaired the meeting. The second one met in Alghero, Sardinia, and this is where the name HGDP was approved. The third meeting was in Pittsburgh and was held in conjunction with anthropologists, thanks to the initiative of professor of anthropology Kenneth Weiss, also an old friend of Cavalli's. Finally, the fourth one was held at the NIH's headquarters in Bethesda, Maryland. This meeting discussed molecular and ethical aspects of the HGDP.

After the first (1991) symposium, Cavalli and other members of what was to become the HGDP urged their colleagues worldwide to collect DNA samples from as many indigenous populations as possible in order to create a universal DNA "bank." This collection, they said, would help us understand the human species and its evolution, and we have seen previously in this book how the study of DNA has helped build phylogenetic trees for the human species. In this type of project, collecting DNA from isolated aboriginal populations is important because these populations have undergone less admixture than other populations. Where does the HGDP fit within the massive efforts of Celera and the Human Genome Institute that culminated in the full sequencing of human DNA?

As Cavalli told us somewhat ironically, the two competing ventures have so far managed to sequence one half of the genome of one man. He was using this comment in a metaphorical sense, but this metaphor has a lot of truth to it. A main goal of the Human Genome Institute (HGI) (and Celera) is to identify the genes believed to be at the root of many diseases; and since the human genome is highly polymorphic, much more than one human genome should be sequenced to identify disease genes. However, to avoid accusations of favoring genes of one individual, both Celera and HGI mixed DNA samples from several individuals, male and female, and sequenced the mixture as if it were DNA from just one individual. Thus, nobody could conclude that any particular sequenced gene belonged to any particular individual. However, Craig Venter, who has since left Celera, has announced that the DNA that was sequenced by his company was mostly his own! What is more, Venter's DNA was isolated from his sperm. Since sperm cells are haploid and only contain a single set of chromosomes, Cavalli could with justification say that we only know half the DNA sequence from that one man. Because of the haploid nature of the sequenced genome(s), it might well happen that these genomes would be lethal in a diploid state. Indeed, it is well known that humans carry several genes that are lethal in a diploid individual. Therefore, there is a chance that the sequenced DNA(s), if diploidized, would correspond to spontaneously aborted or stillborn fetuses! On the other hand, Cavalli pointed out to us that the Venter team's accelerated work pace was very useful because it pressed HGI scientists to work all that much harder than they had before Celera joined the race to sequence the human genome.

For researchers interested in medical and evolutionary genetics, the work of Celera and HGI is only a beginning. Ideally, to apply the knowledge gained from DNA sequencing to medical purposes and use it as a predictor of disease, one should know the DNA sequence of each and every human being on the planet, especially if treatment for a disease depended on the genetic makeup of each individual. This is not yet feasible, but it is likely that certain disease genes are more prevalent in certain subpopulations than in others and, hence, more easily discoverable. Similarly, if one were interested in human evolutionary history, one could use massive sequencing information from isolated populations to refine phylogenetic trees. But here, there is a problem: aboriginal populations are rapidly being absorbed by surrounding populations. Thus, HGDP participants insist that they must act rapidly and collect DNA samples *now*. So we see that both HGI and HGDP must expand their DNA studies in order to reap the benefits of their preliminary investigations.

So far, studying people's DNA for medical and evolutionary genetic purposes seems innocent enough. However, it is already known that some relatively homogeneous populations (such as the Icelanders and the Finns, for example) show either a higher or a lower incidence than the world average for some diseases. Doubtless, aboriginal populations will show the same trend. Thus, the HGDP, while studying human diversity and evolution, is also a potential source of genetic knowledge pertaining to human diseases. This will mean profit for some through the development of therapies against these diseases. Therefore, the HGDP has the potential of being much more than an academic pastime dealing only with human phylogenetic trees. We will see later in this chapter that possible commercial aspects of the HGDP, which were never of any interest to Cavalli, have met with significant opposition.

At its beginning, the HGDP had to iron out the differences that existed between its member scientists. For example, one question was, "How do we define a population?" According to Cavalli, a population should be identified by factors such as language, geography, and endogamy, or marriage within a group. Others, specifically Allan Wilson, favored a "grid" approach in which DNA samples would be collected at regular geographic intervals, such as every fifty kilometers or so. A compromise was reached in which four hundred populations would be surveyed, with blood taken from twenty-five individuals per population. These numbers were of course an initial, tentative suggestion.

There were also problems of cost and logistics. Not only does it cost money for scientists to travel to the remote locations where aboriginal populations live (such as the deserts and jungles of Africa, the jungles of Brazil, Borneo, and the Andaman Islands), it is also expensive to send back to genetic laboratories blood samples preserved well enough to culture white blood cells. Transformed, sometimes called "immortalized," white blood cells were originally used by HGDP scientists as a source of DNA. "Immortalized" is a strange but commonly used word that simply means that these cells can be cultured for a prolonged period of time. The reasoning was that white blood cells could be cultivated for indefinite periods of time in the lab after transformation with Epstein-Barr virus. These immortalized white blood cells could then be used as plentiful sources of DNA for typing. For example, the first Pygmy cell lines used by Cavalli were created in 1984 and 1985 thanks to the crucial help of Judith and Kenneth Kidd of Yale University.

The HGDP project was funded (rather modestly) and work began. But there was another problem: white blood cells (also called lymphoblasts) present in harvested blood survive for only a few days in the test tube, meaning that they had to reach a lab equipped for cell cultivation within that period of time. And indeed, things did not always go smoothly for the HGDP at the beginning. Both Cavalli and Barry Hewlett told us about delayed flights and broken test tubes that hampered their efforts.

Yet all these problems paled by comparison with what was to come. A little over a year into the project, moral and ethical objections to it, and protests against it, began stirring up a storm of controversy that continues to this day. As a result, the HGDP did not survive as originally conceived. As Walter Bodmer told us, insisting first on studying vanishing populations may have been a tactical mistake. According to him, and in hindsight, if the HGDP had focused more on "regular" populations and the value of polymorphisms for medical purposes, much grief could have been avoided. When reviewing our manuscript, Cavalli told us that the present chapter was the most difficult one for him. We consider this is because the HGDP became such a difficult chapter in his life.

The HGDP Controversies

The HGDP has played a significant role in a number of social controversies. As we will see, these are particularly significant within the field of anthropology. These controversies include: (1) the concept of race in the human species, (2) the ethics of taking blood or other bodily substance samples from indigenous human populations ("biocolonialism"), and (3) "identity politics," or the debates over scientific as opposed to cultural constructions of human identities (in terms of ethnicity and ancestry). We address each of these issues in turn.

Race

The term *race* is generally understood to mean a subset of a species exhibiting heritable features that distinguish it from other groups of the same species. As we saw in chapter 1, racial classifications (and racist thinking) were very much a part of nineteenth-century anthropology but were steadily challenged over the twentieth century. Based on a survey by Leonard Lieberman and Rodney Kirk, the majority of anthropologists today believe that biological races do not exist in the human species.[1] Cavalli emphatically agrees with this position and has himself enumerated the scientific grounds on which the concept of race is invalid with respect to humans. Cavalli's position on race (and racism) is clear and consistent in all his writings. His perspective is perhaps best summed up in his statement in his book *The Great Human Diasporas*, that "the confusion, misery, and tragic cruelty caused by [the concept of] racial differences between humans are, to use the words of Shakespeare's Macbeth, 'a tale told by an idiot, full of sound and fury, signifying nothing'" (125).

The painful history of this racist "sound and fury" in the United States can be highlighted with reference to the **eugenics** movement. The concept of eugenics—that is, the selective breeding of human beings for the betterment of the "human race"—was formulated in its modern form by Francis Galton of England in an 1869 book entitled *Hereditary Genius*. In it, Galton ranks the intellectual standard of the ancient Athenians two points above the intellectual standard of the men in his own society. In turn, the latter rank two points above the intellectual standard of the Negroes. Galton's views were enthusiastically espoused in the United States by Charles Davenport in a 1911 book entitled *Heredity in Relation to Eugenics*. By then, Mendel's laws of heredity had been rediscovered, prompting Davenport to advocate limiting the reproduction of feebleminded people and those suffering from venereal diseases (hence

---

1. A few anthropologists argue that while not scientifically precise, racial classifications can be useful and serve worthy ends, as in the fields of forensic anthropology (for example, determining the "race" of a crime victim) and in medicine (determining the susceptibility of a patient to specific diseases). However, the "races" of interest to medical geneticists are not the "classical" ones, i.e., those corresponding to continents. A classical example are Jews, by no means a "race," but rather a cluster of populations sharing a common ancestry which they also share with other people of Middle Eastern origin. Jewish populations separated as a result of their diasporas of 2,600 and 2,000 years ago. The separations gave these populations a chance to diverge genetically, and it is this divergence that is medically important.

Some genetic diseases are found almost exclusively among Ashkenazi Jews (who migrated to East Central Europe and later to North America) but are not found among Sephardim Jews, who are mostly located around the Mediterranean and in South America. Ancestors of Ashkenazi Jews who migrated to Europe about one thousand years ago were a small group whose size grew rapidly. In this way, one or a few recessive mutant individuals multiplied and gave rise to homozygous (affected) individuals. A similar situation is encountered among French Canadians and Afrikaners, whose populations increased about a thousandfold in the past 350 years.

presumed to have innate lewd behavior), in particular. His premise was that both feeblemindedness and uncurbed sexual practices led to a life of crime. Davenport's understanding of Mendel's laws was incorrect, as pointed out at the time by future Nobel Prize winner, geneticist Thomas Hunt Morgan of Columbia University. Nevertheless, in spite of this criticism, several American states established and enforced sterilization laws.

By and large, proponents of the eugenics movement in the United States ranked black people and those of southern European origin below those of northern European origin, with Nordic people at the top. Somewhat similar views are still held today in the United States, as evidenced by Richard J. Hernnstein and Charles Murray's hugely successful book, *The Bell Curve* (1994). Eugenics was not restricted to the United States, however. More notoriously, the eugenics movement of Nazi Germany is very well known. Nazi eugenics not only prescribed sterilization; it also endorsed genocide based on "race."

Still, the American eugenics movement dwindled in the early 1930s. Unfortunately, sterilization and antimiscegenation laws remained on the books in some states for several more decades. For example, the State of Oregon continued to practice forced sterilization of some feebleminded people and "wayward" girls (but not "wayward" boys, of course) until 1981. It was only in November 2002 that Oregon's governor, John Kitzhaber, formally apologized to the surviving victims of this horrible practice. And as we know, some political factions, as well as many individuals in the United States, are opposed to immigration that, they claim, can only result in the weakening of this country.

Cavalli's points against the concept of race are frequently made in the works of other contemporary anthropologists as well. These points include, first, the fact that we cannot sort human beings, either phenotypically or genotypically, into neat, bounded categories. Phenotypically, some have assumed that a category of "black" or "African," for example, can be characterized by a set of traits like dark skin, wiry hair, and thick lips and noses. But each of these traits can be found in combination with the opposite of the others. For example, many people of southern India have dark skin but straight or wavy hair and thin lips and noses. And all over Oceania and Southeast Asia we can see combinations of dark skin with diverse forms of hair and facial features. Let us be clear: *Nowhere do phenotypic traits combine in such a way that we can use a set of traits to clearly and exclusively define any group in contrast to others.* Attempts to define "races" genotypically meet with no greater success. So, for example, individuals of a group one may wish to label as "Native American" do not all share identical sets of genes. A group of individuals as a whole may be characterized by, for example, certain haplotype frequencies, but these haplotypes also exist among other groups.

Second, Cavalli and others have long pointed out that biological variation within groups one might classify as separate "races" is often greater than variation among the groups (see chapter 3). *This is true for virtually all hereditary features. For this reason, any attempt to classify human races is arbitrary.* The attempts that have been made to divide humans into races belie the arbitrary nature of the task since the results have ranged from three to sixty or more human races, as already noted by Darwin. The interest in "pure races" typical of much racism is a phenomenal mistake: "pure" races do not and cannot exist, and ethnically mixed marriages, far from being deleterious, are at the least harmless and may perhaps even be genetically advantageous for the progeny. Of course, racial and social intolerance may make such marriages difficult in some social environments.

Most of the traits people conventionally use to discuss "racial" variation are phenotypic traits like skin color and certain facial features. These traits appear now most probably to be simply responses to climate. Those people settling in sunny, hot regions of the globe may, through natural selection, develop darker skin to protect themselves from the sun's radiation (and hence from skin cancer and burns). For populations at higher latitudes this protection from the sun's rays is less necessary (see chapter 3, note 2). For similar adaptive reasons, populations in cold climates may develop long, thin noses, allowing them to warm the cold air they breathe in before it reaches their lungs.

The idea that superficial human biological differences are responses to climate requires more research for full substantiation. But the significance of this notion and this kind of research for conceptions of race and for racism is quite clear. As Cavalli and his son, Francesco, have written:

> Adaptation to climate for the most part requires changes of the body surface, because this is our interface with the outside world. *It is because they are external that these racial differences strike us so forcibly, and we automatically assume that differences of similar magnitude exist below the surface, in the rest of our genetic makeup. This is simply not so: the remainder of our genetic makeup hardly differs at all.* (*The Great Human Diasporas*, 124; emphasis in original)

Yet another issue confounding racial classification is that most phenotypic traits exist around the world as continuous gradients, defying any attempt to use them as boundary markers between groups. Thus, one can see that skin color gradually darkens as one scans human populations starting from Europe and moving down toward the Equator, although skin color is lighter in Central African forests. Therefore, there are no breaks in the pattern that would allow us to draw clear lines between groups. These gradients are called **clines**. Many anthropologists advocate that a study of human biological variation should use a clinal approach and abandon the concept of race, or subspecies. In this way gradients such as skin color can be studied in terms of human evolutionary adaptations, not in terms of whether or not they can be used to exclusively define human racial categories, as discussed by anthropologist Faye Harrison.[2] Over the last decades, anthropological understanding of human biological variation has been shifting away from a focus on human subspecies to a focus on clines and the evolutionary significance of clinal, not group, variation.

Along with challenging the scientific validity of the concept of race, Cavalli has also made significant contributions to the struggle against racism. Most notable were his attacks, in the early 1970s, on racist arguments that average IQ ("intelligence quotients") differences between blacks and whites in the United States are genetically determined, or innate. At about this same time, a parallel antiracism argument was being developed on the other side of the Atlantic by Cavalli's colleague Sir Walter Bodmer, among others. In England, Bodmer debated the race / IQ relationship promoted by a well-known psychologist, Hans Eysenk, as well as Arthur Jensen, who was then also in England.

As we saw in chapter 4, Cavalli publicly debated both Arthur Jensen and William Shockley on this issue. At that time, Jensen had published his finding that the difference in IQ between American blacks and whites was about fifteen points and his interpretation that this difference was genetic and unchangeable. With Walter Bodmer in 1970, Cavalli demonstrated how this interpretation was unfounded, pointing to several cultural or environmental factors—quality of schools, social discrimination, economic deprivation, and so on—that could easily be seen to account for the IQ differences. In addition, he argued that a truly "culture free" test of intelligence (whatever that is) is impossible.

Since that time some studies of adopted children have shown support for these ideas. One notable study by Sandra Scarr showed that the IQ scores of American black children who had been adopted into affluent white families did not differ significantly from that of white children adopted by these same families. This study also showed that at age seven, both the white and black adopted children had IQ scores above that of the general population, suggesting the influence of advantaged households on children's learning and test performance.

Studies of twins raised in different households and adopted children in "interracial" households may mitigate the impact of clearly racist statements like those of Jensen and Shockley. However, these studies are necessarily rare and difficult to arrange. Further, a test of IQ differences that truly controlled for

---

[2] Another meaning of "race" is a specific lineage within a larger species group, such that members of the lineage only mate within it. There are no human races in this sense of the term (Cavalli-Sforza and Cavalli-Sforza 1995; see also Templeton 1998).

factors such as the impact of social stigma, discrimination, racial stereotyping, and so on is not practically or morally feasible. As anthropologist Marvin Harris stated in his textbook *Culture, People, Nature*:

> In order to control for these factors, we would have to place a sample of black infant twins for adoption, one from each pair in a white household and the other in a black household. Then we would do the same for a sample of white infant twins—half in white households and half in black households. And even then we would have to change the color of the white children to black and the black children to white to control for possible effects of social rejection of transracial fostering. (82)

Interpretations of IQ test scores often appear to be biased by observers' social position. In the 1980s another report emerged showing that Japanese children averaged IQ scores eleven points above American children. As Cavalli has noted in *Genes, Peoples, and Languages*, many Americans in that case began to blame poor American schooling for this difference, whereas they had not blamed poor schooling for blacks after the 1970s reports of a similar IQ difference between black and white children.

Clearly Cavalli has abandoned the "race" concept and fought against racism. As for the HGDP, he and others felt it would generate data that further disputed the concept of biological races among humans because it would show overall genetic similarity among all humans and confirm that genetic variation within groups is greater than that between them.

The HGDP is not based on "racial" classifications; rather it documents variations in gene frequencies among populations, based on geographical location, linguistic affiliation, and other factors. Yet the human phylogenetic trees Cavalli has constructed from genetic data rest on the grouping of human beings into named categories—Asiatic, African, and so forth. This has raised the question, What are *these* groupings if not "races"? In a 1999 article Debra Harry and Jonathan Marks comment on what they see as the mixed messages of the HGDP: "We learn on the one hand, that the HGDP will prove that races do not exist . . . and on the other hand, that its results can be summarized in widely publicized color-coded maps in which 'Africans are yellow, Australians red, [Mongoloids blue] and Caucasoids green' " (304).

What, then, are the HGDP "populations" and in what sense are they not "races"? These populations are units, or categories, devised by Cavalli and his coworkers on the basis of a number of factors. In *The History and Geography of Human Genes* (1994), Cavalli and his colleagues wrote: "The code we eventually adopted for classifying our populations is geographical-anthropological (physical)-linguistic-ethnographic, the order of the four words reflecting the average importance of each criterion in making decisions in uncertain cases" (23).

Using geography, language, and so on to make the classification is a reasonable way to have the categories drawn up to resemble as closely as possible "populations" in the sense of localized groups whose members interbreed. These populations are then found to be distinctive in terms of gene frequencies—all of which are phenotypically invisible and the main value of which is that they can tell us how populations diverged from one another in the past, suggesting probable periods of expansion and paths of human migrations in prehistory.

The HGDP populations are not "subspecies." However they are categories of people showing biological, if invisible, genotypic differences, and for that reason alone the HGDP "populations" become "race"-like in the minds of many. At the same time (as discussed in chapter 3), *some* categorization of people is unavoidable in the study of human population genetics. It is important to remember that the HGDP is not an exercise in how to best divide humanity into discrete groups; rather it *starts* with a classification of people, for convenience and out of necessity. As Cavalli's linguist colleague Merritt Ruhlen wrote in his 1991 book: "Classification, or taxonomy, is a fundamental pursuit of science, an indispensable first step in the search for understanding" (1). In the HGDP, categorization of people is a necessary first step; the end point is to show how and where human groups moved in the past.

But to some, *any* grouping of humans into biologically differentiable units (whether these are called "races" or not) automatically opens the door to racism. Biologically rooted difference potentially spells essential inequality. Feminists have made a similar point with respect to notions of biologically based differences used to account for behavioral and psychological contrasts between (and differential social evaluations of) men and women. Still others argue that while the HGDP is not itself racist, it can promote racism. In a 1999 article Joseph Alper and Jon Beckwith, for example, wrote that, "Concentrating attention on genetic differences among groups has the obvious danger of providing fodder for those who promote racist policies and ideology" (286).

This issue of race has placed Cavalli's work at the center of one of the most troublesome social issues of our time. What began as an attempt to improve our understanding of human diversification and migration is now ensnared in charges and denials of racism. Beneath this issue is a broader one: Can humans even consider biological difference apart from notions of essentialized inequality? Cavalli's work forces us to confront that question, and the HGDP is now a major arena in which this issue is being confronted.

Biocolonialism

Indigenous groups in the United States, New Guinea, and some other countries have protested the taking of bodily substances (such as blood or hair) from indigenous peoples for use in genetic research. Some have referred to this process as "biocolonialism," or "biopiracy," and to the genetic researchers as "gene hunters" or "vampires." In the United States a "Model Resolution" against biocolonialism prepared for use by tribal councils by the Indigenous Peoples Council on Biocolonialism (IPCB) stated in 1999 that "scientific research and genetic exploitation of indigenous peoples represents the greatest threat to American Indians since the European colonization of the Americas."

A number of biocolonial concerns are expressed by this and other Native American organizations. As pointed out by Aroha Mead in 1996, one concern is that blood, hair, and so on are, in many indigenous worldviews, sacred parts of a person; scientific research on these substances is then seen as a violation of indigenous cultural values. Many consider the taking of these substances by outsiders for their own interest or profit as a violation of individual human rights. These days most organizations taking DNA samples are under restrictions to do so only with the "informed consent" of donors. But some people question the extent to which "consent" of donors has been "fully informed," especially in cases where cell samples are "immortalized" for future uses of which the donor may be unaware. This point is made by, among others, Debra Harry, a Native American and outspoken critic of the HGDP. David Resnik has also questioned the meaning of informed consent where the linguistic and cultural barriers between outside researchers and indigenous communities are great. As acknowledged by Henry Greely, there have also been questions about the need for (and problems of legitimately obtaining) group as well as individual consent where indigenous communities are concerned. According to Steve Olson, the idea here is that since the genetic studies focus on groups and not individuals, the group as a whole (represented by some leaders or a council) should have a say in whether or not their members donate DNA samples. The HGDP has adopted a group consent policy for its work in North America.

Another concern is indigenous peoples' feeling that they have been treated as mere research objects, or curiosities, not as equal participants in genetic research projects. Many, such as Joseph Alper and Jon Beckwith, have said that there are few, if any, benefits of this research to indigenous peoples but enormous advantages for genetic researchers and enormous profits to biotechnology and pharmaceutical companies. They also object that the U.S. government is willing to spend millions of dollars to preserve indigenous peoples' DNA, whereas federal money for preserving or improving the lives of indigenous peoples is scarce. Even in the case of so-called "disease genes," where genetic research could potentially benefit humanity in general, or populations within it, critics point out that from the point of view of an indigenous people, money could be better spent on health programs and services.

There are also concerns that DNA research on indigenous peoples might challenge tribal groups' rights to their land or natural resources (for example, through a scientific claim that a particular person or people are not really indigenous). Some individuals—for example, Sandra Awang—and groups have further suggested that genetic research could lead to a new type of genocidal biowarfare—biological weapons targeting groups of people with particular genes. Fortunately this fear is unfounded. As Steve Olson and others, notably in Great Britain, point out, a biological weapon targeting certain genes (if ever developed) would not decimate a population but only certain percentages of people within it and other exposed populations.

In addition to these concerns some indigenous people such as Aroha Mead have criticized the Eurocentric scientific worldview that sees nature as something to exploit and refashion and that seeks to "commodify" all aspects of human life. F.C. Dukepoo pointed out that indigenous peoples have been particularly outraged at the commodifying of life through the patenting of their DNA. Finally, indigenous communities have raised objections to the scientific hegemony of genetics in defining their own identity (as discussed more fully in the next section).

Others claim that some of these criticisms have been wrongly focused on the HGDP. David Resnik has emphasized, for example, that whereas other genetic research has led to patents, the HGDP research has not done so. Those defending the project have also countered that some of the issues of concern to indigenous peoples have come about through other private and even covert genetic research, whereas the HGDP itself offers an organized, publicly accountable, and more responsible way in which genetic research can be conducted. This point has been made by Kenneth Weiss, David Resnik, and Kenneth and Judith Kidd. The HGDP, many—including Cavalli—have pointed out, has carefully considered the ethical issues and developed a moral ethical protocol in 1999. Supporters also emphasize the potentially tremendous scientific and medical benefits of the HGDP. On the issue of informed consent, Kenneth Weiss and others such as H. Cann, Cavalli, and thirty-nine other signatories have insisted that the HGDP is aware of the problems, respects the need for both individual and group consent, has made every effort to secure meaningful consent in an appropriate way, and does not collect samples or conduct any research without this consent.

Scientific Hegemony Over Human Identity Construction

A number of anthropologists have criticized Cavalli's work, especially that connected with the HGDP, for its role in what they consider to be "identity politics." Human identity—notions of who one is, relatedness to others, descent, ancestry, and so on—are, in this view, cultural constructions; a genetic determination of "who's who" is yet another construction, but one that claims scientific authority and hence ultimacy.

Some charge that the HGDP claims a genetic determination of human identity as definitive and imposes this construction on others, hegemonically defining their identity for them. This position is articulated by Jonathan Marks, who explains the title of his 2001 paper, "'We're Going to Tell These People Who They Really Are': Science and Relatedness," as

> derived from the justification given for the Human Genome Diversity Project (HGDP) by a spokesman, to an audience of bioethicists, in an unsuccessful attempt to drum up support for the project at the International Congress of Bioethics in 1996. A Native American activist responded from the audience: "I know who I really am. Shall I tell you who *you* really are?"

Genetic dominance in "identity politics" is, then, basic to other processes whereby scientific constructions can be used to control and exploit others, especially indigenous peoples.

Among some anthropologists there is also the idea that the definitive nature of genetic connections reflects a specific Euro-American cultural conception that "biologizes" individual identity and human relationships. This whole issue can be elucidated in terms of a current trend in anthropological studies of kinship, long a major topic of interest within the field. For decades anthropologists studied the kinship systems of non-European peoples, showing how important kinship was in the social, economic, religious, and political lives of those in small-scale (hunting and gathering) societies and in peasant populations. They drew complicated diagrams showing people's genealogies and systems of kinship terminology. They wrote innumerable books describing various people's modes of descent, residence, and marriage rules that together formed the bedrock of these people's social organization, way of life, and cultural beliefs. In these studies anthropologists assumed that, just like Euro-Americans, people everywhere base their understanding of themselves, their personhood, and their important "kinship" relations with others on their understanding of their own biological connections.

This assumption was later challenged by anthropologist David Schneider. Schneider insisted that the traditional anthropological understanding of kinship was flawed because of its assumption that kinship, or notions of human relatedness, everywhere followed ideas of biological relationship, or shared "biogenetic substance." He pointed to many systems in which other people base their so-called kinship on patterns of food-sharing, common residence, religious rituals, child adoptions, and many other cultural notions and processes.

Interpreting other people's "kinship" on a biogenetic premise is then an ethnocentric imposition of Euro-American ideas on the cultural realities of other peoples. Schneider's ideas are still being debated within anthropology, but they have spread far beyond kinship studies, influencing thinking on other kinds of social identity such as gender and ethnicity. Following Schneider, for example, feminist anthropologists Sylvia Yanagisako and Jane Collier have charged that a construction of gender in terms of male/female differences in biological reproduction is a Western cultural view and not universal. Schneider's ideas became part of the postmodern and, in more extreme forms, antiscience movement within cultural anthropology (as discussed in earlier chapters). It has left a deep impact in anthropology, one suspicious of biologized cultural conceptions and, in some quarters, of biology itself.

As a result of the controversies reviewed here, the HGDP has had a troubled and poorly funded history. A lot of the furor was instigated early on by the Rural Advancement Foundation International (RAFI), based in Canada. This organization works against the patenting of plant genetic material in developing countries by outside corporations. It charges that these corporations use this material to create hybrid seeds that they then sell to developing countries at very high prices. The RAFI considered that projects like the HGDP would do something similar—extract and patent genetic material from indigenous peoples and then use it to make expensive drugs for sale back to them. As mentioned earlier, the HGDP has not patented any genetic material.

In 1993 the RAFI notified the World Council of Indigenous Peoples of meetings planned in connection with the HGDP and itself recommended that all HGDP sample-taking should be ended. Various indigenous organizations were then in communication with one another and began protests against the HGDP, many calling for a halt to the project. Cavalli reported that until this action was taken by the RAFI, negative reaction to the HGDP had been fairly mild. But then, with the RAFI's accusations, the flood gates opened. The whole experience reminded Cavalli of a famous aria from Rossini's Italian opera, *Il Barbiere di Siviglia* (The Barber of Seville): "La calunnia è un venticello" (Calumny is a breeze). In this aria, slander against Count Almaviva, started by a rival, begins as a whisper, a simple rumor. The breeze then gains strength and becomes a devastating storm that almost destroys the victim of the slander.

As the HGDP faced these problems, Cavalli asked Stanford law professor Henry Greely to study the ethical aspects of the HGDP. With his background in bioethics, Greely represented the project to indigenous groups and created the Model Ethical Protocol to govern the collection of DNA samples. Greely met with indigenous groups to discuss the project and allay fears of it, but according to Steve Olson, his efforts were not well received.

Cavalli also reported to us that he was met with hostility in meetings with some indigenous groups. At one point Cavalli contacted a Native American geneticist, Frank Dukepoo, to discuss the issues with him. Dukepoo seemed quite willing, so Cavalli asked him to select a small group of concerned Native Americans to bring with him to Stanford for a constructive discussion. Dukepoo came with Debra Harry and five other women. Cavalli reported that at this meeting the group of Native Americans did not engage in discussion or dialogue but only expressed its hostility toward, and hatred of, the HGDP. Even Dukepoo, who initially had seemed enthusiastic about the meeting, had, according to Cavalli, apparently adopted the position of the others. In contrast to this experience, Cavalli reported that his discussions with Australian aborigines about the HGDP and their participation were rather positive. In light of all this controversy, we discuss the current status and activities of the HGDP in the next chapter.

The HGDP and Anthropological Tensions

As we have seen, the HGDP received criticism from many quarters, from bioethicists and indigenous peoples as well as anthropologists. The HGDP is particularly a point of tension within anthropology precisely because the three controversies reviewed above strike at the very heart of the discipline. Given its own particular history and consequent sense of guilt, the profession of anthropology especially fears three things: charges of racism, the wrath of indigenous groups, and attempts to deny or supersede local peoples' own understanding of themselves and their world. The HGDP has stirred up all of these fears.

In short, the HGDP reminds many anthropologists of what they most dislike about their own professional history. Since the nineteenth century, anthropologists have moved away from race and racism, but they are acutely aware of racism in the history of the field. They are likewise remorsefully aware of anthropologists' own exploitation and mistreatment of indigenous and other studied peoples; the "gene-hunting" HGDP reminds them of how anthropologists themselves have mined native peoples for data from which only anthropologists profit. Finally, anthropologists have increasingly emphasized the importance of understanding and respecting native peoples' own views and definitions of reality. These concerns are no longer exclusive to anthropology but are also moving into the public consciousness, as both anthropological perspectives and the voices of indigenous peoples and other disadvantaged groups impact public awareness.

In retrospect, a definite mistake made by the HGDP was its failure to include representatives from indigenous groups in the very initial planning and design of the project. Doing so might have been costly and may have taken time, but it also may have served a great educational function, avoiding many of the more serious and painful pitfalls and reactions to the HGDP among indigenous groups and anthropologists. Some have charged that the project did not initially consult with anthropologists, but this is not true; anthropologists were involved from the very beginning. In fact in 1993 a symposium was organized to discuss the HGDP's populations and sampling, with some sixty anthropologists in attendance. According to Cavalli, these anthropologists made no objections to the design or methods of the HGDP and gave no warning of how it might be received among indigenous or professional groups or what ethical confrontations might result.

For all the criticism of the HGDP, many anthropologists and others are interested in or committed to a scientific understanding of the human species, its history and variation. Many believe that a scientific exploration of human variation (whether this is variation by population or by gender) need not inevitably end in racism, gender inequality, or exploitation of disadvantaged groups. They consider, on the contrary, that scientific investigation of human global variation will improve our understanding of ourselves as a single species with more commonality than difference. Regardless of one's position on the HGDP, what we can learn from its experience is how quickly and how deeply issues of human biology and human biological variation affect our social lives, evoking questions about ourselves as individuals, groups, and as a species. Cavalli's work has clearly moved to the heart of these questions.

How did Cavalli react to all these criticisms, hostilities, and direct attacks concerning the HGDP? He told us he had learned from Peter Underhill a beautiful American expression that explains it all: "It is easy to recognize a pioneer: his back is full of arrows." In the next and concluding chapter, we summarize the latest developments of the HGDP, reflect upon Cavalli's career, and discuss how his contributions may impact the future.

Chapter 8
The Legacy

Legacy often implies controversy. Without controversy we have simple history: the passage of time, the passing on of genes, or the vertical transmission of stable cultural messages. Legacy is intellectually proactive. It is the affirmation that the future is more than a mere continuation of the past. Legacy is breakthrough: it is revisionism, revolution, or perhaps just a new way of looking at things. In our view, Cavalli's legacy will be his vision that a great deal about human evolution—in its biological, cultural, and linguistic aspects—can be understood through looking at how and where human groups have moved over the earth, preserving more stably those characteristics which are passed from generation to generation through specific conservative mechanisms of transmission. This basic but brilliant idea has a number of ramifications that have been explored throughout this book. Cavalli's vision has brought about a new way of looking at human evolution and, like every novelty, it has been controversial. It has also brought Cavalli's research to the interface between genetics and society.

In this chapter we discuss the past and present impacts of Cavalli's research on science and society at large. In the meantime, Cavalli's peers and others have recognized his work in the form of numerous prizes, society memberships, and honorary degrees. Just to name a few, Cavalli is a member of the National Academy of Sciences (United States), the Royal Society, the French Academy of Sciences, the American Academy of Arts and Sciences, and the Italian Academy of Lincei, and he has honorary doctorates from Cambridge and Columbia universities as well as several Italian universities. He is also a recipient of the prestigious International Balzan Prize, a prize whose monetary reward is of the same magnitude as the Nobel Prize. In October 2003, Cavalli was made an honorary citizen of the city of his birth: Genoa.

Cavalli was also invited to become a member of the Pontifical Academy of Sciences (housed by the Vatican). He hesitated because his understanding of genetic diseases—and their cohort of pain, grief, and suffering afflicting patients and their families—has led him to become a strong supporter of prophylactic termination of pregnancy of embryos doomed to severe diseases. This position clashes with that of the Roman Catholic Church. Raising this concern, Cavalli was told that the academy was only asking him to believe in the truth. He became a member. And as a member of the Pontifical Academy of Sciences he has brought the issue of abortion to the attention of the other academicians—so far with very little success. It should be noted that the Pontifical Academy of Sciences does not restrict membership to a single religion; Muslims and Jews, as well as atheists, can be found among its members.

In the following, we review Cavalli's major accomplishments and analyze some of the follow-ups to them. (See figure 8.1.) Before this discussion of Cavalli's legacy, let us first recapitulate in a roughly chronological fashion what the major achievements of Cavalli and his associates have been:

1. Elucidation of the *F* sexual system in the *Escherichia coli* bacterium
2. Realization of the importance of genetic drift in the history and geography of human genes
3. Key contributions to mathematical genetics, including the methodology and the drawing of the first human phylogenetic tree

Figure 8.1 Cavalli, with author Linda Stone (Portland, Oregon, 2003). (From the authors' collection)

4. Formulation of the demic diffusion model for the spread of agriculture in Neolithic Europe and of population expansions in general
5. First application of principal component analysis of populations using genes, and the drawing of geographic gene maps representing the spread of *Homo sapiens* in space and time
6. Integration of genetics and anthropology to understand human prehistory
7. Establishment of a quantitative theory of cultural evolution and transmission
8. Development of a theory of the coevolution of genes, languages, and culture via mechanisms of both genetic and cultural transmission
9. Founding of the Human Genome Diversity Project
10. Use of Y chromosome haplotypes and microsatellites to trace the evolution and movements of modern humans

Bacterial Genetics

We saw in chapter 2 that Cavalli did not discover bacterial sex; Lederberg and Tatum did. However, Cavalli *explained* important aspects of bacterial sex. This explanation was extremely controversial at the time of its announcement. In his acceptance speech for the Kistler Prize—awarded by the Foundation for the Future—in September 2002, Cavalli said, referring to his scientific activities in the 1950s, "I thus grew accustomed to being considered a little crazy even by colleagues for whom I had respect. But I was somewhat surprised and saddened to discover that even some good scientists are not always interested in novelties, and tend to react very often with skepticism." Cavalli told us he wanted to be remembered for his discovery of the *F* system in *E. coli*, the system that makes sex possible in this organism (remember that the Lederbergs had obtained very similar results, and they and Cavalli published together). Unfortunately, modern genetics textbooks generally credit William Hayes (now deceased) for this discovery. We believe that this textbook phenomenon is akin to mitosis, the process by which one cell divides to produce two identical daughter cells, and so on. Somewhere, there must still exist the "mother of all textbooks," the one that started the rumor that Hayes explained bacterial sex and that Cavalli simply discovered *Hfr*. Authors of subsequent textbooks may have copied and propagated the rumor, without studying the relevant literature. To add insult to injury, Hayes *also* discovered a second, and different, *Hfr*! However, Cavalli insisted both verbally to us and in a 1992 article that Hayes was aware of his results before reaching similar conclusions. This is also Joshua Lederberg's opinion. This textbook phenomenon may have its origin in the fact that Hayes's experimental approach is easier to explain to students than the more complete (and yes, perhaps somewhat rabbinical) story told by Cavalli and the Lederbergs. In the end, Cavalli considers that there never was a real competition between him and William Hayes although initially there were some differences in interpretation that were ironed out by later work.

On the other hand, Cavalli did not get the complete *Hfr* story, as we know it today, quite right. For him (and the Lederbergs), mating between *E. coli* cells was compared to the union between sperm and egg in mammals and plants. There, *all* the genes from the male and female reproductive cells find themselves associated in a one-celled embryo called a zygote. In mammals and plants, the zygotic cell divides, differentiates, and eventually becomes an embryo and then an adult. The adult keeps all the genes the zygote received from both parents. Clearly, in *E. coli*, this does not happen, as Cavalli well knew. There, many genes from one of the parents simply disappear. Thus, Cavalli and the Lederbergs thought that, somehow, genes from that bacterial "zygote" (as they imagined it) were *eliminated*. This putative, and selective, gene elimination took place close to the *Hfr* locus on the *E. coli* chromosome that Cavalli had mapped correctly. This locus was also seen as some type of breaking point (see chapter 2). Bacterial genetics, in its early days, was horrendously complicated, even "Byzantine," as geneticists Gunther Stent and Richard Calendar put it, and experimental results could be interpreted in a number of ways. Well, it turns out that the breaking point hypothesis was right, but the elimination hypothesis was not quite as general.

Hayes had proposed at the same time an alternative explanation to bacterial zygotic gene elimination. For him, a bacterial zygote never formed. On the contrary, the *F+* (or *Hfr*) cells transferred only *some* genes to the *F–* recipient, not all of them. Thus, gene elimination did not need to take place. We know now that in fact an *F+* (or *Hfr*) cell actually *can* transfer *all* of its genes to an *F–* under special circumstances. In that sense, both Cavalli and Hayes were right, except that Hayes captured the more prevalent situation in *E. coli* matings while Cavalli took mating to its ultimate possibilities (is this an Irish versus an Italian worldview?).

Today, we know that the conjugation tube that unites an *F+* (or *Hfr*) cell to its *F–* mate, and through which the *F+* genes travel, is rather fragile. When the tube breaks, only a few genes are transferred (à la Hayes), but when the tube does not break, all the genes are transferred (à la Cavalli). It takes about ninety minutes for an *Hfr* to transfer all its genes to an *F–*. Given that bacterial sex can easily be compared with human sexual intercourse (who can help but make the comparison?), and given that *E. coli* can divide in less than thirty minutes, Jacques Monod, the great French geneticist, once commented that *E. coli*, as opposed to humans, can enjoy bliss for over three times as long as its lifetime. If ever one wanted proof that evolution is not progress, this might be it!

All things considered, what were the implications of *F* and *Hfr*, these two devilishly complex *E. coli* mating elements? There were many. First, *Hfr*s were used to demonstrate the circularity of the *E. coli* chromosome and allowed precise mapping of hundreds of genes on this chromosome. Second, Cavalli's discovery that *Hfr* can revert to regular *F+* (this happens when the integrated *F* factor leaves the chromosome and becomes again a piece of free DNA) was put to use to isolate a special category of *F* factors called *F*'s. An *F'* (that is, an F prime) is generated when integrated *F* leaves the *E. coli* chromosome carrying with it a restricted number of genes that were originally located on the chromosome. This discovery was instrumental in the development of the operon theory, which explains how the expression of bacterial genes is regulated. The discoveries just described were all made by Jacques Monod, François Jacob, and Elie Wollman at the Pasteur Institute in Paris, which for a number of years became the world center for bacterial genetics. In 1965 Jacob and Monod received the Nobel Prize in Physiology or Medicine for these and other contributions to science. Who knows what could have happened if Cavalli had been working then in Paris as opposed to Milan where he was all but isolated?

Years later, many other types of small circular DNA molecules were found in all sorts of bacterial genera. Some of these are called conjugative plasmids, and they all basically work like the *F* factor. Their use considerably advanced the knowledge of the genetics of bacteria other than *E. coli*, including human pathogenic bacteria. Finally, the *F* factor discovered by Cavalli is now used to clone large pieces of DNA containing up to 300,000 base pairs, the equivalent of dozens or more genes. These clones are routinely used in the sequencing of the genomes of complex organisms, including humans.

Then there is the story of polygenes in bacteria. This contribution remains controversial today because bacterial geneticists still cling to the idea that a given phenotype should be under the control of a single gene (or an operon, a genetic unit that acts as a single gene). For example, the system studied by Cavalli in 1951–1952—resistance to the antibiotic chloramphenicol—is commonly taught as being under the control of a single gene. Yet this does not agree very well with the way this antibiotic works, and it does not agree with Cavalli's findings. On the other hand, even though the 4 million or so base-pair genome of *E. coli* was fully sequenced years ago, we still do not understand what roughly 40 percent of this genome does. In other words, we do not know the functions of well over one thousand *E. coli* genes. Are there polygenes hidden somewhere among these unknown genes? What we *do* know, now, is that yeast—not a bacterium but another type of single-celled organism—does host polygenes. As Peter Oefner, one of Cavalli's collaborators and codiscoverer of the yeast polygenes, told us, their efforts to study polygenes would have been very different had not Cavalli with his knowledge of things past, his depth, been present.

Theoretical Biology and Mathematical Genetics

Much of biology has no theoretical foundation. Contrary to physics, biological science in general consists of a large collection of empirical observations unconnected by sets of quantitative paradigms. It may well be that the whole of biology will never lend itself to some great unified mathematical theory that can explain everything. There is one exception in the life sciences, however: genetics. As we saw, genetics can easily be expressed in statistical terms and is thus amenable to mathematical treatment. In the end, genetics can be seen as a unification of two great theories: Mendel's theory of the gene and Darwin's theory of evolution by natural selection. This union is sometimes referred to as neo-Darwinism, and we have seen that Ronald Fisher was one of the creators of this grand unified theory. Cavalli, once he embraced human population genetics, followed in the footsteps of his mentor.

For example, Cavalli and associates made important contributions to the study of human evolution with or without selection, and as driven by random drift. They developed cluster analysis of human populations, applied phylogenetic trees to the human family for the first time, and applied principal component analysis to the study of prehistoric human migrations and to similarities between human populations. They also developed approaches using archaeological information and genetic techniques using mutation rates to allow more precise timing of population divergence. By doing so over several decades, they set down solid theoretical foundations that can be applied not only to human evolution but also to evolutionary science in general.

Today, Cavalli continues to contribute to mathematical genetics. In a study published in the *Proceedings of the National Academy of Sciences* (2004), he has returned to his demic diffusion model, using the wave of advance theory. He asked himself the following question: Using computer simulations, what is the fate of a mutation (a genetic variant) taking place near or on the wave front of an expanding population? One of two things can happen. The genetic variant can stay where it is in the wave of advance and produce little progeny. But if the variant is present at the extreme front of the wave of advance, it has a much greater chance of becoming more frequent and traveling toward the periphery of the expansion, in a sense "surfing" on the wave. As they multiply and expand geographically, the variants have a high chance of constituting 100 percent at the edge of the expansion. This may be a mechanism of punctuated evolution through which expanding human populations diversified as they colonized the world.

These results are in good agreement with the general geography of genes, which shows highest frequencies of variants in extreme locations, far away from the East African origin of modern humans. This type of analysis promises to generate good methods for locating the origins of various mutations in both time and space and—connecting these mutations with their genealogies—might enable investigators to refine their studies of human expansion over the four corners of the planet.

Although the adjective has probably been overused, Cavalli proved one more time that genetics is a holistic life science. It is a great irony that, in the United States, some universities have decided to merge their genetics department with others, such as biochemistry and/or microbiology. The view of the reformers was that the life sciences can now be subdivided into two large categories: the molecular life sciences and the nonmolecular sciences. The result of this thinking was that some genetics departments that traditionally included faculty members more molecularly inclined (doing DNA work and the like), as well as population geneticists who are by definition mathematically inclined, have been split. Many genetics departments are in fact no longer in existence. The molecular geneticists have been grouped with biochemists (who have very little knowledge of and interest in genetics), while the population geneticists have generally been melded with ecologists, physiologists, zoologists, and botanists. This trend is representative of an intellectual attitude that divides scientists into two categories: the splitters, who like to subdivide disciplines, and the lumpers, who prefer to agglomerate them. In places where the splitters prevailed, genetics was doomed. In contrast, Joshua Lederberg understood very well the importance of population genetics when he developed the Genetics Department at Stanford in the late 1960s. In his words to us: "Understanding populations was one of the main distinctions of genetics from molecular biology, and I felt that was vital for a medical school. I wanted to be sure we covered all the bases."

In this book, we saw that an application of genetic principles can perfectly well include both molecular and population aspects. For example, Cavalli has resorted to both approaches to tackle very interesting problems. In a nutshell, it may well be that genetics, the sole life science that is equipped with a theoretical foundation, must be distinguished from other life sciences, such as traditional biochemistry, which basically consists (in its narrowest applications) of a collection of techniques aimed at solving very narrowly defined problems. Cavalli's achievements, and his impact on his junior collaborators, would not have been possible in an environment where splitters establish barriers between subdisciplines. It is to the credit of Stanford University that it promotes multidisciplinary research. We will return to this point later.

Cultural Evolution, Genes, and Archaeology

We saw in chapter 4 that Cavalli and Feldman's theory of cultural evolution was more or less ignored in cultural anthropology. But Cavalli is convinced that anthropology in fact suffers from the lack of research dedicated to cultural transmission. He notes that genetics began with Mendel as the study of genetic transmission. This idea of transmission was sufficiently complicated that no one understood or was interested in it for thirty-five years, from 1865 to 1900. The rebirth of genetics and its expansion was due to another discovery in genetic transmission: the partial invalidity of the law of independence of transmission of different characters, which led to the chromosome theory of biological inheritance. It then took more than fifty years before it became possible to guess correctly the chemical structure of what is inherited. In Cavalli's view, anthropology, like genetics, will advance as a field only when it begins to focus on mechanisms of transmission—in this case cultural transmission.

Cavalli also suggested to us in 2003 that there is a major difference between genetic and cultural evolution that may not have been noted before. In human populations, 84 percent of the genetic variation is common to all groups and only 16 percent is characteristic of a group. This 16 percent is most likely caused by adaptations to different climates, which arose largely after the spread of modern humans to the whole world. This percentage will of course increase as human groups have time to differentiate. In our modern world, however, this increase may be counterbalanced by globalization and some increase in intergroup marriage. Currently we have no estimate of the cultural variation between and within human populations, but Cavalli suggests it is likely to show the reverse trend—namely, that cultural variation within populations will be minor compared to cultural variations between populations. This trend, he says, is clear for languages: a language must be mutually understandable within a population in order for communication to take place. Other cultural traits must similarly remain common and compatible for a group to function. Hence, in contrast to genetic variation, cultural variation will become lower within than between populations. In the case of language, mutual understandability must originally have been within a range of one or a few days' walking time. But over time, nation states have enlarged the range within which one speaks the same national language. Similarly, trade has made common linguafrancas necessary and useful over even wider ranges (e.g., Swahili in East Africa).

Meanwhile, within a recent but now growing area of cultural anthropology—one that welcomes a scientific approach and focuses on evolutionary mechanisms of culture—Cavalli and Feldman's theory may yet flourish. One cultural anthropologist who is very actively bringing Cavalli's ideas to the fore is Cavalli's colleague Barry Hewlett (see chapter 4). Hewlett seeks to integrate Cavalli's ideas on mechanisms of cultural evolution and demic diffusion with some other concepts and approaches in cultural anthropology. He has already begun this work in collaboration with Italian geneticists C. R. Guglielmino and Annalisa de Silvestri. These three in 2002 published a paper that seeks to explain the distribution of particular cultural traits (practices and beliefs) across Africa. Already showing an integration of Cavalli's ideas with cultural anthropology, the authors decide to call these traits "semes," rather than "memes" (see chapter 4). As "seme" derives from "sign," it better conveys the symbolic dimension of cultural beliefs and practices, or the symbolic nature of culture.

Hewlett and his coworkers selected forty-two African ethnic groups for which sufficient genetic, linguistic, and cultural data were available. They then calculated the genetic, geographic, and linguistic distances among these groups and recorded the distribution of semes among them. The semes (109 in all) covered a variety of beliefs and practices that could be grouped into larger categories such as kinship, division of labor by sex, house construction, and sociopolitical organization. They noted which semes were shared among which cultures, asking, Why do cultures share these semes? Performing complicated mathematical procedures on their data, they were able to test for the likely influence of three different theoretical models of seme sharing. These models were (1) cultural diffusion (borrowing semes from neighbors), (2) local adaptations (different peoples develop the same or similar semes as an adaptation to similar environments), and (3) demic diffusion (as in Cavalli's theory, people take their semes with them as they migrate).

The results of their study are very interesting. They found, for example, that demic diffusion accounted for the largest number of shared semes (twenty in all). They also found that many of these semes had to do with kinship, family, and community organization. Cultural diffusion accounted for fewer semes (twelve), many having to do with house construction. Local adaptation accounted for very few (four) shared semes, one being a kinship seme, the existence of small extended families. Their findings will likely surprise many anthropologists who consider that a great deal of culture is due to human ecological adaptations.

Most significantly, this study showed that many of the shared semes explainable by demic diffusion are often considered as core cultural features of sub-Saharan Africa—for example, residentially independent polygynous families and the practice of horticulture. The study suggests, then, that demic diffusion, as originally outlined by Ammerman and Cavalli in their study of the spread of agriculture in Europe, can be a very powerful explanation for the distribution of semes in Africa. In addition, these authors related their findings to Cavalli and Feldman's theory of cultural transmission, noting, for example, that the kinship and family semes are more conservative than other semes, in part because they tend to be vertically transmitted.

Hewlett's study very strongly supports the ideas already implied in Cavalli's view of culture—namely, that a great deal of human cultural history can be understood in terms of how and where humans have moved, taking their genes, languages, and cultures with them and preserving most that which is vertically transmitted. For Hewlett and his Italian colleagues, this work in genes and culture, inspired by Cavalli, has only just begun.

Hewlett's study was multidisciplinary and collaborative, continuing the tradition established by Cavalli himself. We have also noted in earlier chapters that multidisciplinary research flourishes at Stanford University, as seen in Cavalli's work with his local collaborators. Currently, Cavalli has ongoing projects with molecular biologists (Peter Oefner and Ron Davis), an anthropologist (Joanna Mountain), and a mathematical biologist (Marcus Feldman). Also, Cavalli has continually inculcated multidisciplinarity into his graduate students, postdoctoral assistants, and junior collaborators. Sometimes this type of approach can lead to surprising results, as the following example demonstrates.

One day Peter Underhill, senior researcher in Cavalli's laboratory, was approached in a café by Roy King, a faculty member in the Department of Psychiatry and Behavioral Sciences at Stanford. Underhill's reaction was at first guarded because, as he told us, he didn't think he was in need of a "shrink." It turned out that King, in addition to being a mathematician and an M.D. by training, as well as a psychiatrist by profession, was also an enlightened amateur archaeologist interested in Neolithic ceramic pottery and anthropomorphic figurines. King had learned somehow that Underhill and Cavalli had drawn a Y chromosome–based genetic map of Europe consistent with the demic diffusion into Europe of Neolithic Middle Eastern farmers. Being familiar with the geographic distribution of Neolithic artifacts in Europe, King wondered if one could correlate the spread of the Middle Eastern Y chromosome haplotypes with archaeological sites where Neolithic pottery and figurines had been found. Within days, these two researchers had their answer (it was yes), swiftly wrote a paper, and submitted it for publication. It appeared in the archaeological journal *Antiquity* in 2002. This is what they found.

As we saw in chapter 6, four European Y chromosome haplotypes show a distribution gradient consistent with a demic diffusion of Neolithic farmers from the Middle East and Anatolia into Europe. If indeed these farmers brought with them their culture as they were moving into southeastern Europe along the Mediterranean coast, as well as in a northwesterly direction, one would expect to find pottery and figurines distributed along the same axes. King and Underhill did find an excellent overlap between these archaeological sites and relevant Y chromosome haplotypes. The match was not perfect, however. For example, Y haplotype frequencies predicted that Andalusia and the French Basque region should contain figurines whereas none have yet been found. On the other hand, Hungary, Croatia, and the Ukraine, which have a low percentage of relevant Y haplotypes, do contain sites where Neolithic figurines have been found. More research is needed to determine the reasons for these two discrepancies. Cavalli pointed out to us later that the two discrepancies noted by King and Underhill may have a simple explanation: first, principal components show that it is only the northern Mediterranean side of Spain that came under Neolithic genetic influence. Thus, the French Basque region was not entered by the Neolithic farmers, and so they would not have left artifacts there. Second, the Balkans experienced a strong, late immigration of southern Slavs, which may have skewed genetic patterns in a different direction.

Since neither Underhill nor King are professional archaeologists, they tried to convince archaeologist colleagues to give them feedback and, if they wished, to become coauthors on their paper. They did receive positive feedback from several sources, but nobody had the courage to cosign the article with them! The reviewers for the journal *Antiquity* mentioned that this was a paper sure to generate controversy.

In yet another way, one can see that Cavalli's early efforts to merge genetics and anthropology are still bearing fruit several decades later. Jared Diamond of the University of California–Los Angeles and Peter Bellwood of Australian National University have recently advocated extending the Ammerman-Cavalli wave of advance model to the whole world. They hypothesize "that prehistoric agriculture [around the world] dispersed hand-in-hand with human genes and languages [and] that farmers and their culture replace[d] neighboring hunter-gatherers and the latter's culture" (603). In other words, they propose to extend the concepts developed by Ammerman, Cavalli, and Renfrew for demic diffusion in Neolithic Europe to the spread of agriculture, genes, and language in the Americas, Africa, East Asia, and Polynesia. Diamond and Bellwood emphasize the complexity of their hypothesis as well as the massive amount of data that will have to be gathered to examine it. They note that, "To extract reliable conclusions from all this evidence will require comparative research on a worldwide scale within multiple disciplines. It is quite a challenge, but a uniquely fascinating one" (603).

Medicine

Practicing medicine was not Cavalli's chosen path; as we saw earlier, he stopped being a practicing physician right after his internship during World War II. However, he retained his medical skills and so treated many patients during his African expeditions. In particular, he cured many people afflicted with yaws (a destructive skin disease very common among Pygmies), thanks to stocks of penicillin donated by Italian pharmaceutical companies.

Cavalli's involvement in medicine also continued in another way: early in his career he became interested in disease genes. Beginning with schizophrenia, he sought to study the association of inherited diseases with genetic markers in order to detect the genes responsible for these diseases. In 1978, Cavalli attended a seminar held at Stanford where he heard about the use of the RFLP technique to identify future sickle-cell anemia sufferers at the embryonic stage. This turned his attention to the potential use of RFLPs in medicine. Later, in 1982, the NIH's Institute of Neurology asked him to chair a committee set up to determine whether funding for the search for the Huntington's disease gene—using the RFLP technology—should be continued. Back then, the research had not made much progress, and Nancy Wexler, the director of the project, had a conflict of interest, being both project investigator and an NIH official responsible for distributing funding. Cavalli insisted that the project continue, which it did, and soon an RFLP marker inherited at high frequency with the disease was discovered. Soon after, this RFLP marker was used to identify human embryos carrying the dominant Huntington's disease gene.

At the same time, Cavalli and collaborator Mary Claire King (former student of famous Berkeley scientist Allan Wilson) received funding from the NIH to pinpoint with RFLP technology the gene responsible for cystic fibrosis (CF). CF is the most prevalent recessive disease in people of European origin, afflicting one out of every two thousand individuals. Oddly, Cavalli and King had previously been rebuffed by the Cystic Fibrosis Foundation, whose scientific committee predicted that the project would not work. Later, the foundation revised its opinion and invited Cavalli to participate in a number of meetings. But by then the field had become crowded and very competitive. In 1985 a Canadian researcher, Lapchi Tsui, located the CF gene to human chromosome 7 using the RFLP technology.

Cavalli and King then decided to concentrate on another dominant genetic disease: neurofibromatosis. Unfortunately, here too they were beaten by another team, that of Mark Skolnick at the University of Utah. Interestingly, Skolnick had been Cavalli's graduate student at both Stanford and Pavia. Equally interestingly, King and Skolnick later became competitors again in the hunt for breast cancer genes: King proved the existence of a dominant gene that causes cancer in heterozygous older women and in homozygous younger women. This gene was shown by her to be located on chromosome 17. However, Skolnick was the first to pinpoint the exact location of that gene (now called BRCA1, for Breast Cancer One) on chromosome 17. Skolnick then started a company, called Myriad Genetics, that now employs five hundred people.

In addition, Cavalli has lately collaborated on two ambitious medical studies—one dealing with autism, and the other dealing with Wilson's disease. Autism is a neurodevelopmental disease that affects about one child in 2,500. It is characterized by little or no verbal communication, lack of social responsiveness, and obsessive behaviors. Several theories have been advanced about the cause(s) of autism. These theories have invoked behavioral, environmental, dietary, viral, autoimmune, and genetic problems. This long list attests to our ignorance of what lies behind the symptomatology of the disease.

A genetic origin for autism, or at least a partial genetic origin, makes some sense when one considers the recurrence of the disease among siblings, which is 2 to 6 percent. This value is much higher than for chance alone, which is 1/2,500 = 0.04 percent. Also, twin studies show that monozygotic (identical) twins are at a 25-fold higher risk for autism than dizygotic (fraternal) twins if one of the twins develops autism. These two observations indicate that a genetic component for autism is likely. On the other hand, autism cannot be due to a single recessive Mendelian gene, in which case recurrence among siblings would be 25 percent. This suggests that autism may be a disease involving polygenes, unlike sickle-cell anemia, galactosemia, Lesch-Nyhan disease, and cystic fibrosis. The question then is: How many genes are responsible for autism?

In an attempt to answer this question, Cavalli and collaborators (including Stanford psychiatrist Roland Ciaraniello, who died of a heart attack while jogging, and statistical geneticist Neil Risch) analyzed 519 microsatellite polymorphisms in DNA isolated from 147 pairs of affected siblings and their parents. The total number of polymorphisms studied was 160,000, an enormous number. Their conclusion was that some ten to twenty genes may be involved in the development of autism. This is a high number of genes, and in spite of the fact that the full human genome has been sequenced, one should not expect these genes to be pinpointed in the near future, although one can hope that some of the most important ones will be discovered soon. Be that as it may, the best and most extensive study (at this writing) on the possible genetic causes of autism is by Risch and his collaborators (1999). Autism is just one of many polygenic diseases in which many genes are involved. Other such diseases are schizophrenia, manic depressive psychosis, hypertension, and arthritis.

Wilson's disease, a degenerative liver (sometimes also neurological) disease caused by the improper metabolization of copper, has been mapped to two closely linked loci on human chromosome 13. What is puzzling is that the age of onset of the disease ranges between four and forty years. Cavalli's study (with his postdoctoral fellow Anne Bowcock) of Wilson's disease (Bonné-Tamir et al. 1990) suggested that the variation in the age of onset may be due to polygenic effects also. Cavalli and Bowcock actually located a disease gene in a region measuring 300,000 base pairs, but were beaten to the finish line by another team which had benefited from their preliminary results. As Cavalli puts it: "In the game of gene hunting, basically a rat race geared by a system that generates maximum anxiety and a fair amount of duplication of effort and cost, those that do not get there first have no recognition at all."

Finally, Cavalli's collaborator, Walter Bodmer, made the important discovery that the variant frequencies of HLAs (human leukocyte antigens) vary enormously from one population to another. Since these HLA variants determine whether transplanted tissues (heart and liver, for example) will be rejected or accepted, it is clear that a good understanding of population genetics is critical in human medicine.

Science, Travels, and Education

Cavalli is a globe-trotter. He told us, perhaps with a touch of bombast, that the only places on earth he had never visited were the North and South Poles. (He also stated that he had no intention of visiting them, ever.) On the other hand, China is a place that Cavalli has visited various times, and he has several collaborators there. One important contribution to our understanding of human origins in China was made by a group under the leadership of Li Jin, a Chinese scientist who was a postdoctoral fellow in Cavalli's lab (see Ke 2001). After a *Homo erectus* fossil (named Peking Man) was discovered near Beijing decades ago, many scientists held the belief that modern Chinese people were descended directly from their own *H. erectus* root. Later on, the presence of *H. erectus* in China was used to support the multiregional hypothesis, which implies that modern Chinese populations could not have descended from the first modern humans who appeared in East Africa. In a massive study conducted in 2001, Li Jin and coworkers typed the Y chromosomes of 12,217 men from 163 populations across Southeast Asia, Oceania, East Asia, Siberia, and Central Asia and concluded that "the data do not support even a minimal *in situ* hominid contribution in the origin of anatomically modern humans in East Asia." Indeed, all tested individuals show the presence of a Y chromosome polymorphism that originated in Africa about 35,000 to 89,000 years ago. These dates are incompatible with the multiregional hypothesis, which implies that *H. erectus's* diversification started occurring about 1 million years ago. Therefore, like all extant human populations, the Chinese also descend from our common African *Homo sapiens* root.

Cavalli was not a coauthor on this article, but his interest in China had germinated many years before this report came out. In the early 1980s, after he had discovered the power of surnames as genetic markers in Italy, he became interested in extending this research to China, the most populated and one of the most literate countries in the world, where, he suspected, surnames were very ancient. At that time, his friend and leading population geneticist James Crow of the University of Wisconsin in Madison had been invited to spend some time in China. Cavalli asked him to try to put him in touch with a Chinese population geneticist. As a result, Cavalli was able to strike up a collaboration with professor Du Ruofu of the Chinese Academy in Beijing. This collaboration resulted in several trips across the Pacific for both of them, as well as for several of Du's students. Cavalli and his Chinese collaborators were able to obtain a sample of 500,000 surnames (a very large sample indeed), which had been collected in the 1981 population census. Interestingly, the Han majority has only 1,063 surnames in existence today. The reason for this small number is that surnames are very old, and many of them probably originated during the Neolithic. Given enough time, provided no new surnames appear (by a process equivalent to gene mutation), drift could reduce the number of surnames to just one. This is not as preposterous as it sounds: in many large Chinese villages and towns everybody has the same surname. Graciously, the Chinese government participated in the research by helping with the computer filing of data. This was a very welcome helping hand, as the Reagan administration was then in the process of cutting funding for science.

Using the methodology he had developed for Italian surnames, Cavalli and coworkers observed profound differences between northern and southern China. In the North, the geographic distribution of surnames was very homogeneous, but in the South it was at least three times more variable. This was very much in line with what one of Cavalli's colleagues at Stanford—Arthur Wolf, an anthropologist who specializes in China—later told him: northern China has one culture, southern China has many cultures. Subsequently, on a tour of southern China, Cavalli realized why: the rugged geography of southern China explains both the genetic and cultural differences. The surname data showed that one could divide China into three parts: the North, the South, and an eastern central region located around Shanghai, somewhat more similar to the South than to the North. The three regions correlate rather well with the history of the Chinese Neolithic, which is almost as old, and presumably independent from, that of the Fertile Crescent and Mexico. According to some paleoanthropologists, differences between northern and southern China already existed in the Paleolithic.

Cavalli and Du then extended their investigations to the question of ethnic minorities in China. The Chinese government recognizes fifty-five such minorities, representing about 100 million people, versus the 1.1 billion people who represent the Han majority. Du had studied genetically thirty-seven of these minorities, and it turned out that his results were in very good agreement with the surname data. The synthesis of their results showed that there has been considerable genetic exchange over time between the Han and the ethnic minorities. It is also known that declaration of one's ethnic minority status considerably fluctuates over time, depending on the advantages or disadvantages of belonging to either an ethnic minority or the Han majority. About half of the minorities are in the southern province of Yunnan, the rest being spread around the country, but mostly near the borders. All minorities have their own language and cultural peculiarities, of which they are very proud.

In the late 1990s American sinologist Victor Mair observed in a museum in Urumqi, the capital of the Chinese province of Xinjiang several recently excavated mummies dated at up to 3,800 years ago. These mummies were well preserved thanks to the dry climate of the desert in which they were found, looked like they were of northern European origin, and wore clothes resembling those made in Austria at that time. Puzzled by these observations, Mair asked Cavalli to help him test the DNA of the mummies and accompany him to China. Not feeling too well at the time, Cavalli was convinced by his wife not to go. Instead, he sent his student, Paolo Francalacci, who is also an archaeologist, with the goal of obtaining mitochondrial DNA samples from the mummies. Francalacci was allowed to take back samples from two mummies. DNA typing showed that one of them was clearly of European origin. Interestingly, the western minorities who live currently in Xinjiang province (the Uygur)—and who display great external morphological diversity—received a 30 to 40 percent genetic contribution from ancestors of European origin (almost certainly speakers of Indo-European languages, Tocharians A and B) whereas the residual 60 to 70 percent of their genes were obtained from admixture with neighbors. It will be fascinating to discover why and how these Europeans undertook this long trek to the confines of China.

Let us now look at what Cavalli is doing in his native country of Italy. There, as in the United States, student interest in the hard sciences is low. To try to remedy this situation, Cavalli and his son Francesco in 2003 published a science book in four volumes for 11- to 14-year-old high school students, explaining how science and scientists work. These books are based on Cavalli's own personal experiences. Hopefully these books will rekindle the enthusiasm of young Italians in the study of the sciences.

Cavalli's concern for the state of science and mathematics in Italy is not new. In 1991 he gave a paper (which was read for him because he was sick at the time) about this concern at a meeting of the Unione Matematica Italiana (Italian Mathematics Union). In it, Cavalli deplored the way that mathematics is taught in a dry, boring fashion to medical students and life sciences students. Remember that Cavalli refused to study engineering, contrary to the wish of his parents. The reason for that was in his words: "È abbastanza ironico che presi questa decisione proprio per timore dell'analisi" ("It is rather ironic that I made this decision because of my fear of calculus"). Evidently, Cavalli overcame that fear later on, and in his paper (published in 1992), he provided some recipes to make mathematics more palatable to students.

Why, he asks, is mathematics for the life sciences taught in such a dry, irrelevant way? Why is it taught in a deductive manner, in which theorems are demonstrated and equations derived? Why not teach math in an inductive manner instead? By using quantitative examples relevant to the life sciences and medicine, students would see the power of mathematics applied to their major subject, would enjoy the beauty of math, and, subsequently, would naturally see how the equations they just used are derived. What is more, they might even be able to remember what they learned! As examples, he cites exponential functions that are used to calculate radioactive decay and the production of mutants by exposure to ionizing radiation, and the logistic growth function. Other examples he proposes are the binomial, Gaussian and Poisson distributions used in studying variation in biological and other phenomena, as well as mathematical modeling with computers to solve numerically problems too complex to be handled mathematically.

The Human Genome Diversity Project

In spite of all the controversy that surrounded it at the beginning, the HGDP continues. In April 2002, Cavalli and forty other academics from the United States, South America, Europe, Africa, and Asia published a letter in the journal *Science* in which they proposed to the international scientific community a worldwide collaboration for the DNA typing of lymphoblasts isolated from fifty-one different populations. Eleven years after the first call for a survey of human genetic diversity, the HGDP delivered. As of October 2003, fifty-three international laboratories had requested DNA samples from the Centre d'Etude du Polymorphisme Humain (CEPH), the Paris institution which keeps the cell collection. Interestingly, in the spirit of collaboration that has always characterized Cavalli, the letter states that, "Finally, cooperative research will be facilitated among researchers who have hitherto been forced into a competitive mode of interaction." Oddly, as of August 2002, Cavalli's lab was the only one among the collaborating laboratories that had not yet received DNA samples from CEPH! This was due to the fact that requests from other laboratories were so numerous that the available DNA was rapidly consumed.

Since cells from 1,064 people were harvested and cultivated by HGDP collaborators, the full DNA typing of all these donors will take time, and it will be a few years before the world community becomes apprised of the discoveries resulting from the HGDP. Some members of the HGDP consider that its medical implications are more important than other aspects. In that respect, it is noteworthy that the prestigious British medical journal the *Lancet* chose as best article published in 2002 a DNA typing study performed with the collection gathered by the HGDP. The typings were done by James Weber's laboratory in Wisconsin, analyzed by Marcus Feldman's lab at Stanford, and published in *Science* (Rosenberg et al. 2002). Cavalli chose not to collaborate but, as a proponent of the HDGP, he is extremely proud that the *Lancet* singled out this article, thereby confirming the medical importance of the idea on which the HGDP was based.

Science and History

One of Cavalli's colleagues at Stanford University told us that if Cavalli had been more focused professionally, he would have secured more and bigger grants. This is not say that Cavalli's research was not well supported by a variety of granting agencies. His colleague was, rather, referring to financial orders of magnitude. Why is it that Cavalli could have received so much more funding than he did? Our answer is that he never agreed to focus on narrowly defined problems. Worse, from the short-sighted viewpoint of many a potential grantor, he bridged the gap between the sciences and the humanities by establishing correlations between languages, genes, archaeology, and cultural anthropology.

Nonetheless, being both an experimenter and a historian, Cavalli sees a difference between the natural sciences and history. In the experimental sciences, Cavalli points out, an investigator can repeat an experiment as often as he or she wants in order to check the reproducibility of results. This is obviously impossible in history since the arrow of time flies in only one direction; even if they were witnessed by (or under the control of) an investigator, past events and sets of conditions can never be reproduced exactly. This holds true not only for the history of human events, as history is traditionally understood; it also holds true for the natural sciences, where time past is of the essence, as in the study of past evolution by natural selection, as well as in the study of the origins of life and the cosmos. However, Cavalli insists, some kind of experimental repetition can be achieved through the use of different disciplines in the study of a phenomenon like human evolution. If the results obtained by the application of different disciplinary approaches converge, one can be reasonably sure that a model based on multidisciplinary empirical discoveries represents historical reality. This is what Cavalli has consistently done, not only in his study of human prehistory but also in his investigations of sociocultural evolution. As we saw, Cavalli has used a combination of phylogeny (genetics and demography), linguistics (including glottochronology and other methods of historical linguistics), mathematics (including advanced statistical theory), and anthropology (including archaeology) to retrace the human past.

Cavalli as a Controversial Icon

We mentioned at the beginning of this chapter that legacy is controversy. Initially, one might think that controversy in the academic arena is a polite affair resting entirely on empirical evidence. In other words, disagreements in academia should be dispassionate and resolved in an entirely objective manner. Well, this idyllic representation is often wrong. Academics can turn into vicious monsters when their pet theories or hypotheses are contradicted by colleagues. Cavalli, with his synthesis of genetics, linguistics, cultural and physical anthropology, and given his long career, was sure to accumulate a significant number of professional enemies. We saw in chapter 7 that the HGDP was accused of being essentially racist. We believe that Cavalli and others have adequately addressed this accusation. But then, other detractors (one of whom privately called Cavalli a "disdainful, ardent self-promoter") imply that if he is not a racist now, he may have been one in the past, simply because he was born and raised in fascist Italy. Cavalli emphatically denies this. This type of *ad hominem* argumentation is political rather than scientific. We address this accusation below, using scientific information published by others, not just easily dismissed hearsay. We also note here that the view of Cavalli as an ardent self-promoter is not widely shared, and particularly not among colleagues who have worked closely with or under him.

In the context of "race" as a sociopolitical issue, we present to the reader a sentence from *Where Do We Come From?* (2002) by population geneticists Jan Klein and Naoyuki Takahata. On page 387, they write in approval that "[Cavalli was first to show in 1966 that] of the total genetic variation observed in the human species, less than 15 percent accounts for differences between races. Cavalli-Sforza apparently found this observation unremarkable, or at least unworthy of a comment." In other words, Cavalli was presenting data on human genetic variation without suggesting implications for the sociopolitical concept of race. They go on to say that another researcher (who confirmed Cavalli's observation) did make a big deal out of this finding six years later, claiming it showed there is no significant genetic difference between so-called human races. While that conclusion may be socially laudable, that researcher, Klein and Takahata point out, failed to specify what amount of genetic difference *would* be sufficient to warrant human racial groupings. This only opened the door for others to claim that the documented genetic difference between human groups *was* sufficient to justify the concept of human races. And so the arguments waged on, with the whole issue muddled and blown out of proportion. Klein and Takahata conclude: "By mixing science with politics, geneticists and anthropologists are committing the same infraction of which they are accusing other scientists, whom they themselves label as racist" (390).

Cavalli's view on races is simple: he thinks it is fair to say that they do not exist, since no sharp racial boundaries can be found anywhere. On the contrary, one finds a continuous and regular increase of genetic distance—a gradation—between two populations as the geographic distance between them increases. The greatest geographic distances are found between continents, and so are the greatest genetic distances. But even so, continents are very far from being genetically homogeneous. Also, since human populations diversified from a small East African population in the last 60,000 years, only small genetic differences could accumulate in such an evolutionarily short time. Some of these differences seem important to some people because they affect easily visible phenotypes such as skin color and body size. However, as far as we know, these adaptations to climate are due to changes in very few genes and do not justify the categorization of human beings into discrete races. Cavalli has explained this view in several of his books and, in his words to us, "Racism is a social disease derived from psychological and social maladaptations that it is a moral duty of everybody to fight, irrespective of their nationality, origin, or religion."

But the race issue is not all that has made Cavalli controversial or has drawn detractors. Cavalli has also been accused of summarily dismissing a seemingly perfectly valid scientific proposition—the multiregional hypothesis. As we saw earlier, there exist two competing hypotheses for the diversification of *H. sapiens* over time and space. One, the uniregional hypothesis, states that modern humans evolved in Africa from *H. erectus* over 100,000 years ago and subsequently migrated and populated the earth as genuine *H. sapiens*. Cavalli is a strong uniregionalist. The uniregional hypothesis recognizes the fact that *H. erectus* migrated out of Africa (as evidenced by the fossil record), and then became extinct without descendants on all the other continents. The other hypothesis, the multiregional hypothesis (also called the Wolpoff-Thorne hypothesis after its authors), also states that *H. erectus* evolved in Africa and started migrating to Europe, Australia, and Asia well over 1 million years ago. But then, in contrast to the uniregionalists, the multiregionalists claim that various populations of modern humans evolved locally in Africa, Europe, Australia, and Asia from their respective branches of *H. erectus*. Multiregionalists also propose that there was significant gene flow (gene exchange) between the African, European, and Asian subpopulations as they were evolving. In other words, multiregionalists posit that *H. sapiens*–specific traits spread to all populations while at the same time regional traits were maintained. Natural selection is also hypothesized to have played a role. So far, the multiregional hypothesis has not been supported by genetic data. Rather, it relies on scanty paleontological evidence, such as provided by the shape of a few fossilized skulls. As one can see, there are irreconcilable differences between the two hypotheses. Presently, the DNA evidence supports the uniregional hypothesis and does not agree well with the multiregional hypothesis (as we saw above in the case of *H. sapiens* in China).

Nevertheless, multiregionalists are not giving up, because more data are necessary to completely rule out their hypothesis. This is why they must feel rubbed the wrong way by Cavalli and associates (and other uniregionalists, who are presently riding the crest of the wave). One major problem is that the multiregional hypothesis, contrary to the uniregional hypothesis, is not genetically precise. For example, it is unclear what the extent and direction of gene flow might have been and what traits were retained (or selected against) by natural selection in the various subpopulations. Indeed, the multiregional hypothesis posits the existence of regionally advantageous mutations that diversified subpopulations from one another, as well as globally advantageous mutations shared by all subpopulations, which made them all distinct from *H. erectus*. Selectively neutral mutations are also assumed to have played a role in this evolution.

One dubious advantage of the multiregional hypothesis, however, is that its imprecision allows it to evolve, absorb new data, and perhaps avoid death thanks to its plasticity. In general, scientists dislike hypotheses that adapt to various circumstances and do not make precise predictions. Proponents of such hypotheses often resort to *ad hominem* arguments such as questioning the ability of their opponents to understand what their vague hypotheses really mean. No wonder then that a frontal attack, supported by real empirical evidence, on marginally substantiated hypotheses is seen by them as arrogant, self-serving, and disdainful.

In 2001 a Japanese team under the direction of Naoyuki Takahata set out to determine under which sets of parameters the multiregional hypothesis could agree with empirical data. For this, they built a computer program which allowed them to simulate the fate of globally advantageous, regionally advantageous, and neutral mutations. They then compared the results of their simulation with actual gene-frequency data. Their observations showed that out of ten polymorphic loci, nine placed the most recent common ancestor of *H. sapiens* in Africa (the tenth locus may not have been statistically significant). Mitochondrial DNA polymorphisms also supported Africa as being the place of origin of our most recent common ancestor. These conclusions are very difficult to reconcile with the multiregional hypothesis, but they are entirely consistent with the uniregional hypothesis. Indeed, for multiregionalists, the most recent ancestor of Europeans should be located in Europe, that of Asians in Asia, and that of Australians in Australia. Of course, a computer program can never prove that the multiregional hypothesis is wrong; it can only show that the latter is not supported by a simulation. At least, however, this is not an encouraging observation for the multiregionalists.

Finally, in 2003 a team headed by Tim White of the University of California at Berkeley published the discovery in Ethiopia of three near-modern human skulls dated at 160,000 years ago. These skulls show modern traits, such as the shape of the braincase, as well as some archaic ones, such as a thick browridge, suggesting that the fossils represent a transition between earlier African hominids and modern *H. sapiens*. The dating of these fossils agrees extremely well with the uniregional hypothesis, which states that modern *H. sapiens* appeared in Africa between 150,000 and 100,000 years ago.

At present the minority and more controversial position in this evolutionary debate is not represented by Cavalli but rather by the multiregionalists. On the other hand, in matters of linguistics, Cavalli has espoused a minority view, that proposed by Joseph Greenberg and his mentee Merritt Ruhlen (see chapter 5). Although their classification of human languages has gained wider acceptance, this classification has been criticized by many others as excessive "lumping" of languages together, particularly with respect to Native American languages. Similarly, Ruhlen's interest in the idea of a single original human language from which other languages descended has been a minority concern.

As we saw, this idea was first proposed more than a century ago by Alfredo Trombetti but was rejected by practically all his contemporaries. However, in addition to Ruhlen, many Russian linguists, as well as Nobel Laureate for Physics Murray Gell-Mann (the discoverer of quarks, who also has remarkable linguistic knowledge), are currently showing strong interest in the notion of a single ancestral language.

Cavalli's Future Endeavors

Cavalli presently directs an NIH-supported project on human DNA variation. Ongoing for nearly a quarter of a century since 1970, it is slated for competitive five-year renewal in 2005. Cavalli has asked his Stanford collaborator of more than thirty-two years, Marcus Feldman, to assume the directorship of the project at that time. However, Cavalli plans to continue collaborating with the Stanford team even after retiring from his directorship. Also, since 1992, Cavalli has been spending six months at Stanford and the rest of the year in Italy, where he has worked on several projects, including science books for junior high school students and a book entitled *Consanguinity, Inbreeding, and Drift in Italy*, published by Princeton University Press in 2004. Last but not least, in line with his interest in culture, Cavalli has begun the direction of an ambitious project to be entitled "History of Italian Culture," a research book in ten volumes. Hopefully, at some point the book will also present a genome survey of the whole country.

Science as Adventure

Scientists are sometimes depicted as aloof high priests of the modern age. In the popular imagination, they manipulate atoms and genes that nobody has ever seen (not true, of course), and when doing this they use a jargon that nobody understands (again, not completely true). Sometimes their discoveries inspire fear and distrust in the public, i.e., when translated into technologies such as nuclear power stations and genetically engineered food plants. At other times, scientists provide us with useful knowledge that allows us to resist disease and live longer, more productive lives. But these applications, if they are possible, almost always come later.

Oftentimes, scientific discoveries have no immediate application, but they tell us something about the universe and about ourselves. In many ways, science is a quest for knowledge alone. Academic science in particular—armed with the tools of deductive and inductive logic—is driven by curiosity. In that sense, science, and by extension, all acquisition of knowledge, is an adventure, an exploration without immediate or practical benefit. Perhaps Cavalli himself summed it up best when of his own career he remarked: "Even then [in Italy during World War II] I had enough freedom to exercise my curiosity. I believe curiosity is an essential ingredient." In the end, Cavalli's curiosity led to an exciting odyssey, an adventure in genes, geography, and human culture that offers the human species a better and deeper understanding of itself.

Glossary

admixture: *See* migration.

allele: *See* gene variant.

autosome: a chromosome that is not sex-specific. Humans possess 22 pairs of autosomes (for a total of 44 autosomes) shared by both sexes, plus two *sex chromosomes* (see below).

base: a component of DNA coming in four varieties: adenine (A), guanine (G), thymine (T), and cytosine (C). In the DNA double helix, A pairs up with T and G with C (also, *see* nucleotide).

Beringia: a stretch of land, now submerged, that once existed as a land bridge between Siberia and Alaska thousands of years ago; it is thought that people from Asia entered the New World through this land bridge during the Paleolithic period or simply walked along the coast of Beringia. Traveling by boat has not been excluded.

bottleneck: a large decrease of the number of individuals in a population. Can be caused by outmigration of a small fraction of a population or by population reduction through accidental circumstances (flood, fire, drought, famine, wars, epidemics, genocide, etc.).

carbon-14 dating: a method of dating organic remains by measuring the amount of radioactivity remaining in them. This dating method relies on the fact that living organisms absorb radiocarbon from the air and that this radioactivity decays at a known rate, halving every 5,700 years.

chromosome: a microscopic cellular structure visible in dividing cells. Chromosomes harbor all of the cell's genes in their DNA except for a very small amount contained in a mini-chromosome present in mitochondria.

cline: a gradient in gene frequencies over geographic space.

Clovis point: a distinctive fluted stone point found at hundreds of archaeological sites in North America, dating from about 11,800 years ago, and thought by many archaeologists to represent the "Clovis people," early inhabitants of the New World.

cognate: in linguistics, words in different languages that are similar enough to suggest common ancestry among the languages.

conjugation: bacterial sex.

consanguineous: a relationship between persons based on descent; pertaining to "blood" kin.

Cro-Magnon: humans who migrated to and inhabited Europe around 40,000 years ago. Cro-Magnon is a locality in SW France where 23,000-year-old skeletons of modern humans were found.

crossing over: a natural breakage and reunion mechanism taking place between two chromosomes. Also *see* recombination.

cultural diffusion: the geographic spread of cultural traits or innovations from one group to neighboring peoples; contrasts with *demic diffusion*.

cultural relativism: the contention that (1) each culture can only be understood on its own terms, or from within; and (2) no culture is superior to another, or that all values are relative to cultural contexts.

demic diffusion: The geographic spread of people during an expansion, or of cultural traits or innovations through the demographic expansion of peoples into areas surrounding the origin of expansion; contrasts with *cultural diffusion*.

DHPLC: denaturing high-performance liquid chromatography. A technique allowing the quick detection of single base-pair differences between two DNA molecules.

DNA: deoxyribonucleic acid. The chemical material of which genes are composed.

drift (also, random genetic drift): The change in frequency of any polymorphic gene (or DNA base-sequence modification) between one generation and the next due to random distribution of the existing forms of the gene in the progeny. If a population goes through a bottleneck followed by major demographic growth, the change in gene frequency is likely to become fixed in that population.

dual transmission theory: a theoretical approach that seeks to understand human evolution through both genetic and cultural transmission.

enzyme: biological catalyst, usually a protein, which facilitates metabolic chemical reactions in living cells. Thousands of enzymes have been isolated and characterized.

Escherichia coli: a bacterium. The most studied living organism to date. Some say that geneticists have two loves: the organism(s) they study, and *E. coli*. Some prefer to add the fruit fly, *Drosophila melanogaster*.

ethnocentrism: the view that one's own cultural or ethnic group is superior to or more natural than another or to others generally; the tendency to interpret ideas or practices of other cultures in terms of one's own cultural categories or values.

ethnography: a description of a particular culture.

eugenics: a program to improve a "race" or "races," especially in humans, through the control of reproduction. Term introduced by Francis Galton.

expansion: prolonged demographic movement of a population followed by its spreading through short-range migrations, as in the case of the demic diffusion of agriculture. Usually due to innovation which improves quality or quantity of food in a culturally transmissible way. Differs from simple migration, which may leave the place of origin empty.

F+: an *E. coli* cell with the ability to donate its gene to a recipient cell.

F–: an *E. coli* cell able to receive genes from an *F+*.

fascism: a system of government; a far-right dictatorship merging state, a nationalistic philosophy, and business leadership. Benito Mussolini's Italy, Adolf Hitler's Germany, Ante Pavelic's Croatia, Miklós Horthy's Hungary during World War II, and General Augusto Pinochet's Chile more recently were fascist states.

gene: the unit of heredity. Genes are made of DNA. Gene names and abbreviations are always italicized.

gene flow: genetic consequence of migration from neighboring population(s), especially if continued for a long period.

gene variant: technically called an *allele*. Different individuals in populations (such as human populations) all share the same set of gene variants though in different proportions. A gene is made of thousands of nucleotides and the same gene in different individuals may show a constellation of variations at each nucleotide location. Gene variants can be identified at the level of the proteins they code for or that of DNA. Differences among variants exist because of different mutations having taken place in different individuals' ancestors.

genome: the suite of all the DNA and genes present in a given species. For example, humans have about 25,000 genes whereas the fruit fly has about 16,000 and the bacterium *E. coli* about 4,000.

genotype: the suite of all gene variants (alleles) in an organism. All humans share the same genome, but the sum total of the variants of the genes present in our genome (our genotype) is unique to each individual, except in the case of monozygotic twins. A genotype is often referred to as a particular combination of alleles, such as AA, Aa, or aa.

glottochronology: a method of dating how long ago two or more languages separated from one another through calculating the proportion of cognate words the contemporary languages share.

haplotype: a particular mutation or set of mutations present in a given stretch of rarely recombining DNA shared by a population. Clusters of Y chromosome haplotypes are called haplogroups. These haplotypes descend from a common ancestor in which mutations arose. They differ among themselves by mutations that occurred later in different lineages.

Hfr: an *E. coli* mutant first discovered by Cavalli-Sforza. These mutants introduce their DNA at very high frequency into *F–* recipient cells.

horizontal transmission: the transmission of cultural beliefs, practices, or knowledge between or among persons of the same generation or where the relative ages and the kinship relation of the transmitter(s) and the recipient(s) is irrelevant.

kinship: The recognition of a special relationship between persons based on, for example, common descent, marriage, particular rituals or other culturally recognized processes.

Kurgan culture: an ancient culture of those people living in an area north of the Black Sea, beginning about 5,000 to 5,500 years ago, distinctive for the domestication of the horse, a pastoral economy, and war chariots.

meme: As suggested by Richard Dawkins, any transmissible unit of cultural information (*see also* seme).

Mesolithic: middle stone age; pertaining to the period of time in human prehistory from the last Ice Age (about 15,000 years ago) to the appearance of agriculture (about 10,000 years ago).

microsatellite: a DNA sequence composed of small blocks of bases that repeat themselves many times.

migration: a movement of individuals to join another population, or of a population to a new location, or to form a new colony away from the mother population. Gene flow, usually repeated over generations, is the migration of individuals from one population to another. Admixture is the result of gene flow or mixing of populations.

mitochondrion (pl., mitochondria): a cellular organelle found in the cytoplasm, whose function it is to metabolize oxygen and produce energy. Mitochondria have their own DNA, which in humans is inherited from the mother.

mutagen: an agent able to change, add, or remove base pairs in DNA and, hence, causing mutations. Examples of mutagens are certain chemicals that react with DNA, as well as ultraviolet light and ionizing radiation.

mutant: an organism having undergone a change in its DNA base pairs. Some mutations have phenotypic effects.

mutation: The most common ones involve the replacement of one or several base pairs in DNA. Others, such as deletions, insertions, and duplications may have more important effects on the DNA sequence. Mutations can either be deleterious, neutral, or beneficial.

natural selection: differential reproduction of genotypes caused by pressure imposed by natural forces like climate, presence of predators, food availability, etc. It is essentially a demographic phenomenon in which some individuals show different death or fertility rates. The composition of a population changes over generations as a result of these differences. The higher capacity of some individuals to survive and reproduce is due to their physiological adaptation to the physical and social environment in which they live.

Neolithic: new stone age; pertaining to a period starting about 10,000 years B.P. in the Middle East and later elsewhere; it is characterized by agriculture and the use of advanced stone tools. Covers different periods in different areas, depending on the time of arrival of agriculture.

Neanderthal: a now extinct species of humans (*H. sapiens*) that lived in Europe and portions of the Middle East between 500,000 and 35,000 years ago. (The more extreme characteristics associated with Neanderthals in Europe appear about 300,000 years ago.)

nucleotide: a DNA base linked to a sugar molecule itself linked to a phosphate group. Nucleotides are the building blocks of DNA.

oblique transmission: transmission of cultural beliefs, practices, or knowledge from persons of a higher generation to those of a lower generation where the transmitters are not the biological or social parents of the receivers.

Paleolithic: old stone age; pertaining to the period of time in human prehisory from the beginning of stone tools (about 2 million years ago) to the end of the last Ice Age (about 15,000 years ago).

phenotype: all the characteristics displayed by an individual, such as flower color in plants, height in humans, fur length in cats, cystic fibrosis symptoms in humans, coat color in horses, etc. Many phenotypic characteristics, but not all, are under the control of genes. The relative importance of genes and environment in the determination of phenotypes varies enormously for different traits.

phylogeny: the science of categorizing, or classifying, living and fossil bacteria, plants, and animals. Phylogeny also defines the temporal and genealogical relationships between life forms, determining which organisms have evolved from which other ones and when. Phylogeny is different from *taxonomy* (see below).

polygenic (adj.): a polygenic trait is a phenotypic trait under the control of several genes. In humans, some traits are monogenic (such as hemophilia) and some are polygenic (such as type I diabetes).

polygyny: the marriage of a man to two or more women at the same time.

polymerase chain reaction (PCR): a technique allowing the production of large amounts of specific DNA fragments from very small samples of DNA.

polymorphism: the property of a given gene and/or its protein product to exist in the form of several variants (alleles) in a population, provided the frequency of this variant is at least 1 percent. The simplest and commonest polymorphism of DNA is the single nucleotide polymorphism (SNP), in which one nucleotide is replaced by another one—for example, a G replacing an A.

postmodernism: in social science, an intellectual movement that advocated epistemological relativity and cultural constructivism, advancing its points often through the blurring of boundaries between conventional categories of thought.

principal component: an element of the statistical method called *principal component analysis* (see following entry). Principal components break down the analyzed data into multiple categories, the first category (first principal component) containing the largest portion of the information gathered and reduced to a single dimension. The second category (second principal component) contains less information than the first, and so forth. This method "mines" data by extracting major independent patterns, estimating their relative importance based on the proportion of the variation they explain, and making it clear which part of the information can be neglected or pursued by other means.

principal component analysis: a statistical technique that extracts information stepwise from a large number of data points. This type of analysis "mines" data by extracting major regularities and estimating their relative importance.

protein: a type of large molecule found in living cells. All proteins are made of amino acids linked together in a linear fashion.

protolanguage: an earlier language from which later, related languages descended over time.

recombination: the ability of two chromosomes to line up, break at a certain point, and exchange DNA among themselves. Recombination is a perfectly natural phenomenon.

restriction enzyme: a protein enzyme which cuts double-stranded DNA at a specific base-pair sequence, and at that sequence only.

seme: a term proposed by anthropologist Barry Hewlett and others to replace "meme," since it better conveys the symbolic nature of cultural phenomena (*see* meme).

sex chromosome: chromosome that is sex-specific. In all mammals, including humans, the female sex has one pair of XX chromosomes in addition to 22 pairs of autosomes (see above). Males, on the other hand, have one X and one Y sex chromosomes, plus the regular 22 pairs of autosomes. All humans inherit one X chromosome from their mother. Sons inherit a Y chromosome from their father whereas daughters inherit a second X chromosome from their father.

sociobiology: the study of the relationship between genes and social behavior; the theory, advanced by E. O. Wilson, that a great deal of animal behavior, including that in humans, is under genetic control.

taxonomy: the practical science of categorizing or classifying living and fossil bacteria, plants, and animals. The possibility of unifying taxonomy and phylogeny is a late aspiration, although not desired by all.

transformation: the ability of some organisms to naturally incorporate free DNA present in their environment. Transformation can be achieved in the laboratory using a variety of techniques. Transformation is at the root of the science of genetic engineering.

tree (also, phylogenetic or evolutionary tree): a tree-like diagram linking organisms and showing evolutionary relationships between these organisms.

vertical transmission: transmission of cultural beliefs, practices, or knowledge from biological or social parents to their children.

wave of advance: a theory which allows the calculation of the rate of spread in space and time of an element such as a gene, a population, an infectious disease, a rumor, a joke, and so on.

References

1. Science and Society, Genes and Culture

Cavalli-Sforza, L. L. and M. W. Feldman. 2003. The application of molecular genetic approaches to the study of human evolution. *Nature Genetics Supplement* 33:266–75.

Harris, M. 1997. *Culture, People, Nature: An Introduction to General Anthropology*. 7th ed. New York: Longman.

Hwang, W. S., Y. J. Ryu, J. H. Park, E. S. Park, E. G. Lee, J. M. Koo, H. Y. Jeon, B. C. Lee, S. K. Kang, C. Ahn, J. H. Hwang, K. Y. Park, J. B. Cibelli, and S. Y. Moon. 2004. Evidence of a pluripotent human embryonic stem cell line derived from a cloned blastocyst. *Science* 303:1669–1674.

Krogh, D. 2002. *Biology: A Guide to the Natural World*. 2d ed. Upper Saddle River, N.J.: Prentice-Hall.

Lurquin, P. F. 2002. *High-Tech Harvest: Understanding Genetically Modified Food Plants*. Boulder, Colo.: Westview.

——. 2003. *The Origins of Life and the Universe*. New York: Columbia University Press.

Omoto, C. K. and P. F. Lurquin. 2004. *Genes and DNA: A Beginner's Guide to the Science and Its Applications*. New York: Columbia University Press.

Ruhlen, M. 1991. *A Guide to the World's Languages*. Vol. 1: *Classification*. Stanford, Calif.: Stanford University Press.

Stone, L. 2001. Theoretical implications of new directions in anthropological kinship. In Stone, L., ed., *New Directions in Anthropological Kinship*, 1–20. Lanham, Md.: Rowan and Littlefield.

2. From Medicine to Bacterial Genetics (1943–1960)

Avery, O. T., C. M. McLeod, and M. McCarty. 1944. Studies on the chemical nature of the substance inducing transformation of pneumococcal types: Induction of transformation by a desoxyribonucleic acid fraction isolated from Pneumococcus type III. *Journal of Experimental Medicine* 79:137–58.

Bodmer, W. 2003. R. A. Fisher, statistician and geneticist extraordinary: A personal view. *International Journal of Epidemiology* 32:938–42.

Cavalli, L. L. and H. Heslot. 1949. Recombination in bacteria: Outcrossing *Escherichia coli* K12. *Nature* 164:1057–1058.

Cavalli, L. L., J. Lederberg, and E. M. Lederberg. 1953. An infective factor controlling sex compatibility in *Bacterium coli. Journal of General Microbiology* 8:89–103.

Cavalli, L. L. and G. A. Maccacaro. 1950. Chloromycetin resistance in *E. coli*, a case of quantitative inheritance in bacteria. *Nature* 166:91–93.

——. 1952. Polygenic inheritance in the bacterium *Escherichia coli. Heredity* 6:311–31.

Cavalli, L. L. and G. Magni. 1947. Methods of analysing the virulence of bacteria and viruses for genetical purposes. *Heredity* 1:127–32.

Cavalli, L. L. and N. Visconti di Modrone. 1948a. Variazioni di resistenza agli agenti mutageni in *Bacterium coli.* I: Raggie ultravioletto. *Ricerche Scientifiche* 18:3–7.

——. 1948b. Variazioni di resistenza agli agenti mutageni in *Bacterium coli.* II: Azotoiprite. 18:1569–74.

Cavalli-Sforza, L. L. 1950. La sessualità nei batteri. *Bolletino Istituto Sieroterapico Milano* 29:281–89.

Cavalli-Sforza, L. L. and J. L. Jinks. 1956. Studies on the genetic system of *Escherichia coli* K-12. *Journal of Genetics* 54:87–112.

Cavalli-Sforza, L. L. and J. Lederberg. 1956. Isolation of pre-adaptive mutants in bacteria by sib selection. *Genetics* 41:367–81.

Hayes, W. 1953. Observations on a transmissible agent determining sexual differentiation in *Bacterium coli. Journal of General Microbiology* 8:72–88.

Jacob, F. and E. L. Wollman. 1961. *Sexuality and the Genetics of Bacteria*. New York: Academic Press.

Lederberg, J. and E. L. Tatum. 1946. Gene recombination in *Escherichia coli. Nature.* 158:558–59.

Lederberg, J., L. L. Cavalli and E. M. Lederberg. 1952. Sex compatibility in *Escherichia coli. Genetics* 37:720–30.

Scammell, M. 1984. *Solzhenitsyn: A Biography*. New York: Norton.

Stent, G. S. and R. Calendar. 1978. 2d ed. *Molecular Genetics: An introductory Narrative.* San Francisco: W. H. Freeman.

Watson, J. D. 1968. *The Double Helix*. New York: Atheneum.

Watson, J. D. and W. Hayes. 1953. Genetic exchange in *Escherichia coli* K-12: Evidence for three linkage groups. *Proceedings of the National Academy of Science (USA)* 39:416–26.

3. The Shift to Human Populations (1952–1970)

Bodley, J. H. 1990. *Victims of Progress*. Mountain View, Calif.: Mayfield.

Bodmer, W. F. and L. L. Cavalli-Sforza. 1976. *Genetics, Evolution, and Man*. San Francisco: W. H. Freeman.

Cavalli-Sforza, L. L. 1971. Pygmies, an example of hunters-gatherers, and genetic consequences for man of domestication of plants and animals. In *Human Genet-*

ics, ch. 3, 79–95. Proceedings of the Fourth International Congress of Human Genetics (Paris, September 6–11, 1971). Amsterdam: Excerpta Medica.

——. 1986. *African Pygmies*. Orlando, Fla.: Academic Press.

——. 2000. *Genes, Peoples, and Languages*. Translated by Mark Seielstad. New York: North Point Press.

Cavalli-Sforza, L. L., I. Barrai, and A. W. F. Edwards. 1964. Analysis of human evolution under random genetic drift. *Cold Spring Harbor Symposium in Quantitative Biology* 29:9–20.

Cavalli-Sforza, L. L. and W. F. Bodmer. 1999 (rpt.). *The Genetics of Human Populations* (1971). Mineola, N.Y.: Dover.

Cavalli Sforza, L. L. and F. Cavalli-Sforza. 1995. *The Great Human Diasporas: The History and Diversity of Evolution*. Translated by Sarah Thorne. Cambridge, Mass.: Perseus.

Cavalli-Sforza, L. L. and A. W. F. Edwards. 1964. Analysis of human evolution. In Cavalli-Sforza and Edwards, eds., *Genetics Today*, 924–27. Oxford: Pergamon.

Cavalli-Sforza, L. L., P. Menozzi, and A. Piazza. 1994. *The History and Geography of Human Genes*. Princeton: Princeton University Press.

Edwards, A. W. F. and L. L. Cavalli-Sforza. 1965. A method for cluster analysis. *Biometrics* 21:362–75.

Hardy, G. H. 1908. Mendelian proportions in a mixed population. *Science* 28:49–50.

Hewlett, B. S. 1991. *Intimate Fathers: The Nature and Context of Aka Pygmy Paternal Infant Care*. Ann Arbor: University of Michigan Press.

Kimura, M. 1968. Evolutionary rate at the molecular level. *Nature* 217:624–26.

Nei, M. and A. K. Roychoudhury. 1972. Gene differences between Caucasian, Negro, and Japanese populations. *Science* 177:434–35.

Sokal, R. R., N. L. Oden, and B. A. Thomson. 1999. A problem with synthetic maps. *Human Biology* 71:1–13.

Wright, S. 1943. Isolation by distance. *Genetics* 28:114–38.

4. Excursions into Human Culture (1970–)

Ammerman, A. J. and L. L. Cavalli-Sforza. 1973. A population model for the diffusion of early farming in Europe. In C. Renfrew, ed., *The Explanation of Culture Change*, 343–57. London: Gerald Duckworth.

——. 1984. *The Neolithic Transition and the Genetics of Populations in Europe*. Princeton: Princeton University Press.

Bodmer, W. F. and L. L. Cavalli-Sforza. 1970. Intelligence and race. *Scientific American* 223:19–29.

——. 1976. *Genetics, Evolution, and Man*. (See above, ch. 3.)

Boyd, R. and P. J. Richerson. 1985. *Culture and the Evolutionary Process*. Chicago: Chicago University Press.

Bruner, E. 1956. Cultural transmission and culture change. *Southwestern Journal of Anthropology* 12:191–99.

Cavalli-Sforza, L. L. 1971. Use of models: Similarities and dissimilarities of sociocultural and biological evolution. In F. R. Hodson, D. G. Kendall, and P. Tautu, eds., *Mathematics in the Archaeological and Historical Sciences*, 535–41. Edinburgh: Edinburgh University Press.

——. 2000. *Genes, Peoples, and Languages*. (See above, ch. 3.)

Cavalli-Sforza, L. L. and F. Cavalli-Sforza. 1995. *The Great Human Diasporas: The History and Diversity of Evolution*. (See above, ch. 3.)

Cavalli-Sforza, L. L. and M. W. Feldman. 1981. *Cultural Transmission and Evolution: A Quantitative Approach*. Princeton: Princeton University Press.

Cavalli-Sforza, L. L. and E. Minch. 1997. Paleolithic and Neolithic lineages in the European mitochondrial gene pool. *American Journal of Human Genetics* 61:247–51.

Durham, W. H. 1991. *Coevolution: Genes, Culture, and Human Diversity*. Stanford, Calif.: Stanford University Press.

Feldman. M. W, L. L. Cavalli-Sforza, and L. A. Zhivotovsky. 1994. On the complexity of cultural transmission and evolution. In G. Cowen, D. Pines, and D. Meltzer, eds., *Complexity: Metaphors, Models, and Reality*, 48–64. Salinas, Calif.: Addison-Wesley.

Fisher, R. A. 1937. The wave of advance of advantageous genes. *Annals of Eugenics* 7:355–60.

Guglielmino, C. R., C. Viganotti, B. S. Hewlett, and L. L. Cavalli-Sforza. 1995. Cultural variation in Africa: The role of mechanisms of transmission and adoption. *Proceedings of the National Academy of Sciences* 92:7585–89.

Harris, M. 1997. *Culture, People, Nature: An Introduction to General Anthropology*. (See above, ch. 1.).

Hewlett, B. S. 2001. Neoevolutionary approaches to human kinship. In L. Stone, ed., *New Directions in Anthropological Kinship*, 93–108. Lanham, Md.: Rowman and Littlefield.

Hewlett, B. S. and L. L. Cavalli-Sforza. 1986. Cultural transmission among the Aka pygmies. *American Anthropologist* 88:922–34.

Hewlett, B. S., A. D. Silvestri, and C. R. Guglielmino. 2002. Semes and genes in Africa. *Current Anthropology* 43:313–21.

Kendall, D. G. 1965. Mathematical models of the spread of infection. In *Conference on Mathematics and Computer Science in Biology and Medicine*, 213–25. London: Medical Research Council.

Kroeber, A. L. 1952. *The Nature of Culture*. Chicago: University of Chicago Press.

Lev-Yadun, S., A. Gopher, and S. Abbot. 2000. The cradle of agriculture. *Science* 228:1602–1603.

Menozzi, P., A. Piazza, and L. L. Cavalli-Sforza. 1978. Synthetic maps of human gene frequencies in Europe. *Science* 201:786–92.

Renfrew, C. 1987. *Archaeology and Language: The Puzzle of Indo-European Origins*. New York: Cambridge University Press; London: Jonathan Cape.

Richards, M., H. Côrte-Real, P. Forster, V. Macaulay, H. Wilkenson-Herbots, A. Demaine, S. Papiha, R. Hedges, H. J. Bandelt, and B. Sykes. 1996. Paleolithic and Neolithic lineages in the European mitochondrial gene pool. *American Journal of Human Genetics* 59:185–203.

Salzman, P. C. 2001. *Understanding Culture: An Introduction to Anthropological Theory*. Prospect Heights, Ill.: Waveland.

Skellam, J. 1951. Random dispersal in theoretical populations. *Biometrika* 38:196–218.
Sokal, R.R. and P. Menozzi. 1982. Spatial autocorrelations of HLA frequencies in Europe support demic diffusion of farmers. .*American Naturalist* 119:1–17.
Sykes, B. 2001. *The Seven Daughters of Eve: The Science That Reveals Our Genetic Ancestry.* New York: Norton.

5. Genes, Languages, and Human Prehistory (1970–)

Bateman, R., I. Goddard, R. O'Grady, V.A. Funk, R. Mooi, W.J. Kress, and P. Cannell. 1990. Speaking of forked tongues: The feasibility of reconciling human phylogeny and the history of language. *Current Anthropology* 31.1:1–13.
Cann, R.L., M. Stoneking, and A.C. Wilson. 1987. Mitochondrial DNA and human evolution. *Nature* 325:31–36.
Cavalli-Sforza, L.L. 2000. *Genes, Peoples, and Languages.* (See above, ch. 3.)
——. 1997. Genetic and cultural diversity. *Journal of Anthropological Research* 53:383–403.
Cavalli-Sforza, L.L., P. Menozzi, and A. Piazza. 1994. *The History and Geography* of *Human Genes.* (See above, ch. 3.)
Cavalli-Sforza, L.L., A. Piazza, P. Menozzi, and J. Mountain. 1992. Coevolution of genes and languages revisited. *Proceedings of the National Academy of Sciences USA* 86:5620–24.
——. 1990. Comment of Bateman et al. 1990. *Current Anthropology* 31:16–18.
——. 1988. Reconstruction of human evolution: Bringing together genetic, archaeological, and linguistic data. *Proceedings of the National Academy of Sciences USA* 85:6002–6006.
Coon, C. 1954. *The Story of Man.* New York: Knopf.
Darwin, C. 1968 (rpt.) *On the Origin of Species* (1859). London: Penguin.
Dillehay, T.D. and M.B. Collins. 1988. Early cultural evidence from Monte Verde in Chile. *Nature* 332:150–52.
Dolgopolsky, A. M. 1988. The Indo-European homeland and lexical contacts of Proto-Indo-European with other languages. *Mediterranean Language Review* 3:7–31.
Gimbutas, M . 1970. Proto-Indo-European culture: The Kurgan culture during the fifth, fourth, and third millennia. In G. Cardona, H.M. Hoenigswald, and A. Senn, eds., *Indo-European and Indo-Europeans,* 155–195. Philadelphia: University of Pennsylvania Press.
Greenberg, J.H. 1987. *Languages in the Americas.* Stanford, Calif.: Stanford Univesity Press.
Guidon, N. and G. Delbrias. 1986. Carbon-14 dates point to man in the Americas 32,000 years ago. *Nature* 321:769–71.
Issac, G. 1976. Stages of cultural elaboration in the Pleistocene: Possible archaeological indicators of the development of language capabilities. In S. R. Harnald, H. D. Steklis, and J. Lancaster, eds., *Origins an-' Evolution of Language and Speech.* Vol. 280:275–88. New York: New York A .demy of Sciences.
Lynch, T.F. 1990. Glacial-age man in South America? A critical review. *American Antiquity* 55.1:12–36.

Marshall, E. 2001. Pre-Clovis sites fight for acceptance. *Science* 291:1730–32.

Nei, M. and A. K. Roychoudhury. 1974. Genetic variation within and between the three major races of man, Caucasoids, Negroids, and Mongoloids. *American Journal of Human Genetics* 26:421–23.

O'Grady, R.T., I. Goddard, R.M. Bateman, W.A. Dimichelle, V.A. Funk, J. Kress, R. Mooi, and P.F. Cannall. 1989. Genes and tongues. *Science* 243: 1651.

Renfrew, C. 1994. World linguistic diversity. *Scientific American* 270:116–23.

——. 1987. *Archaeology and Language: The Puzzle of Indo-European Origins.* (See above, ch. 4.)

Ruhlen, M. 1991. *A Guide to the World's Languages.* Vol. 1: *Classification.* (See above, ch. 1.)

Tobias, P.V. 1988. Evidence for the early beginnings of spoken language. *Cambridge Archaeology Journal* 8:72–94.

Wolpoff, M.H. 1989. Multiregional evolution: The fossil alternative to Eden. In P. Mellars and C.B. Stringer, eds., *The Human Revolution: Behavorial and Biological Perspectives of the Origin of Modern Humans.* Vol. 1. Edinburgh: Edinburgh University Press.

6. On to DNA Polymorphisms and the Y Chromosome (1984–)

Ammerman, A.J. and L.L. Cavalli-Sforza. 1984. *The Neolithic Transition and the Genetics of Populations in Europe.* (See above, ch. 4.)

Bowcock, A. M., J.R. Kidd, J.L. Mountain, J.M. Hebert, L. Carotenuto, K.K. Kidd, and L.L. Cavalli-Sforza. 1991. Drift, admixture, and selection in human evolution: A study with DNA polymorphisms. *Proceedings of the National Academy of Sciences USA* 88:839–43.

Chikhi, L., R.A. Nichols, G. Barbujani, and M.A. Beaumont. 2002. Y genetic data support the Neolithic demic diffusion model. *Proceedings of the National Academy of Sciences USA* 99:11008–13.

Hammer, M.F., A.B. Spurdle, T. Karafter, M.R. Bonner, E.T. Wood, A. Novelletto, P. Malaspina, R.J. Horai, T. Jenkins, and S.L. Zegura. 1997. The geographic distribution of human Y chromosomes. *Genetics* 145:787–815.

Hammer, M. F. and S. L. Zegura. 1996. The role of the Y chromosome in human evolutionary studies. *Evolutionary Anthropology* 5:116–134.

Piazza, A., S. Rendine, G. Zei, A. Moroni, and L.L. Cavalli-Sforza. 1987. Migration rates of human populations from surname distributions. *Nature* 329:714–16.

Semino, O., G. Passarino, P.J. Oefner, A.A. Lin, S. Arbuzova, L.E. Beckman, G. De Benedictis, P. Francalacci, A. Kouvatsi, S. Limborska, M. Marcikiæ, A. Mika, B. Mika, D. Primorac, A.S. Santachiara-Benerecetti, L.L. Cavalli-Sforza, and P.A. Underhill. 2000. The genetic legacy of Paleolithic *Homo sapiens sapiens* in extant Europeans: A Y chromosome perspective. *Science* 290:1155–59.

Sgaramella-Zonta, L. and L. L. Cavalli-Sforza. 1973. A method for the detection of a demic cline. In N. E. Morton, ed., *Genetic Structure of Populations*, 128–35. Honolulu: University of Hawaii Press.

Sykes, B. 2001. *The Seven Daughters of Eve: The Science That Reveals Our Genetic Ancestry*. (See above, ch. 4.)

Thomas, M. G., K. Skorecki, H. Ben-Ami, T. Parfist, N. Bradman, and D. B. Goldstein. 2000. Origins of Old Testament priests. *Nature* 394:138–40.

Underhill, P. A., L. Jin, R. Zemans, P. J. Oefner, and L. L. Cavalli-Sforza. 1996. A pre-Columbian Y chromosome–specific transition and its implications for human evolutionary history. *Proceedings of the National Academy of Sciences USA* 93:196–200.

Underhill, P. A., L. Jin. A. A. Lin, S. Q. Mehdi, T. Jenkins, D. Vollrath, R. W. Davis, L. L. Cavalli-Sforza, and P. J. Oefner. 1997. Detection of numerous Y chromosome biallelic polymorphisms by denaturing high-performance liquid chromatography. *Genome Research* 7:996–1005.

Underhill, P. A., G. Passarino, A. A. Lin, P. Shen, M. Mirazón Lahr, R. A. Foley, P. J. Oefner, and L. L. Cavalli-Sforza. 2001. The phylogeography of Y chromosome binary haplotypes and the origins of modern human populations. *Annals of Human Genetics* 65:43–62.

Zei, G., A. Piazza, A. Moroni, and L. L. Cavalli-Sforza. 1986. Surnames in Sardinia. III. The spatial distribution of surnames for testing neutrality of genes. *Annals of Human Genetics* 50:169–80.

7. The Human Genome Diversity Project (1991–)

Alper, J. S. and J. Beckwith. 1999. Racism: A central problem for the Human Genome Diversity Project. *Politics and Life Sciences* 18:285–340.

Awang, S. 2000. Indigenous nations and the Human Genome Diversity Project. In G. J. S. Dei, B. L. Hall, and D. G. Rosenberg, eds., *Indigenous Knowledge in Global Contexts: Multiple Readings of Our World*, 120–36. Toronto: University of Toronto Press.

Bodmer, W. F. and L. L. Cavalli- Sforza. 1970. Intelligence and race. (See above, ch. 4.)

Cann, H. M. (plus 39 additional authors) and L. L. Cavalli-Sforza. 2002. A human genome diversity cell line panel. *Science* 296:261.

Cavalli-Sforza, L. L. 1998. The Chinese Human Genome Diversity Project. *Proceedings of the National Academy of Sciences USA* 95: 11501–503.

——. 2000. *Genes, Peoples, and Languages*. (See above, ch. 3.)

Cavalli-Sforza, L. L. and and F. Cavalli-Sforza. 1995. *The Great Human Diasporas: The History and Diversity of Evolution*. (See above, ch. 3.)

Cavalli-Sforza, L. L., P. Menozzi, and A. Piazza. 1994. *The History and Geography of Human Genes*. (See above, ch. 3.)

Cavalli-Sforza, L. L., A. C. Wilson, C. R. Cantor, R. M. Cook-Deegan, and M. C. King. 1991. Call for a worldwide survey of human genetic diversity: A vanishing opportunity for the Human Genome Diversity Project. *Genomics* 11:490–91.

Davenport, C. B. 1911. *Heredity in Relation to Eugenics*. New York: Holt.

Dukepoo, F. C. 1998. Commentary of "F. Jackson": An American Indian perspective. *Science and Engineering Ethics* 4:171–80.

——. 1999. It's more than the HGDP. *Politics and Life Sciences* 18:293–97.

Galton, F. 1979 (rpt.). *Hereditary Genius* (1869). London: Julian Friedmann.

Greely, H.T. 1998. Legal, ethical, and social issues in human genome research. *Annual Review of Anthropology* 27:473–502.

Harris, M. 1997. *Culture, People, Nature: An Introduction to General Anthropology.* (See above, ch. 1.)

Harrison, F.V. 1998. Introduction: Expanding the discourse on "race." *American Anthropologist* 100.3:609–31.

Harry, D. 1998. Tribes meet to discuss genetic colonialization: A report from the Colonialism Through Biopiracy Conference (*see* www.ipcb.org).

——. 2001. Biopiracy and globalization: Indigenous peoples face a new wave of colonialism (*see* www.ipcb.org).

Harry, D. and J. Marks. 1999. Human population genetics versus the HGDP. *Politics and the Life Sciences* 18.2:303–305.

Hernnstein, R.J. and C. Murray. 1994. *The Bell Curve.* New York: Simon and Schuster.

HGDP (Human Genome Diversity Project). 1999. *See* www.stanford.edu/group/morrinst/hgdp/faq.html.

IPCB (Indigenous Peoples Council on Biocolonialism). 1999. Model Resolution (*see* www.ipbc.org).

——. 2000. Indigenous peoples, genes, and genetics (*see* www.ipbc.org).

Kidd, K.K. and J.R. Kidd. 1999. Evidence of preliminary results in human genome diversity research. *Politics and Life Sciences* 18:314–16.

Lieberman, L. and R.C. Kirk. 2001. "Race" in anthropology in the 20th century: The decline and fall of a core concept (*see* www.chsbs.cmich.edu/rod_kirk/norace/tables.htm).

Marshall, E. 1998. DNA studies challenge the meaning of race. *Science* 282:654–55.

Marks, J. 1995. *Human Biodiversity: Genes, Race, and History.* New York: Aldine De Gruyter.

——. 2001. "We're going to tell these people who they really are": Science and relatedness. In S. Franklin and S. McKinnon, eds., *Relative Values: Reconfiguring Kinship Studies*, 355–83. Durham, N.C.: Duke University Press.

Mead, A. 1996. Genealogy, sacredness, and the commodities market. *Cultural Survival Quarterly* 20:46–53.

Olson, S. 2001. The genetic archaeology of race. *Atlantic Monthly* (Spring 2001): 69–80.

Resnik, D.B. 1999. The Human Genome Diversity Project: Ethical problems and solutions. *Politics and Life Sciences* 18:15–23.

Ruhlen, M. 1991. *A Guide to the World's Languages.* Vol. 1: *Classification.* (See above, ch. 1.)

Scarr, S. and R.A. Weinberg. 1976. I.Q. test performance of black children adopted by white families. *American Psychologist* 31:726–39.

Schneider, D.M. 1984. *A Critique of the Study of Kinship.* Ann Arbor: University of Michigan Press.

Templeton, A.R. 1998. Human races: A genetic and evolutionary perspective. *American Anthropologist* 100.3:632–50.

Weiss, K.M. 1996. Biological diversity is inherent in humanity. *Cultural Survival Quarterly* 20:26–29.

——. 1999. Legitimate and illegitimate views of the Human Genome Diversity Project. *Politics and Life Sciences* 19:334–35.

Yanagisako, S.J. and J.F. Collier. 1987. Toward a unified analysis of gender and kinship. In J.F. Collier and S.J. Yanagisako, eds., *Gender and Kinship: Essays Toward a Unified Analysis*, 14–50. Stanford, Calif.: Stanford University Press.

8. The Legacy

Bodmer, W. F. and L.L. Cavalli-Sforza. 1976. *Genetics, Evolution, and Man.* (See above, ch. 3.)

Bonné-Tamir, B. M. Frydoman, M. S. Agger, R. Bekeer, A. M. Bowcock, J. M. Hebert, L. L. Cavalli-Sforza, and L. A. Farrer. 1990. Wilson's disease in Israel: A genetic and epidemiological study. *Annals of Human Genetics* 54:155–68.

Cann, H.M., C. de Toma, L. Cazes, M.-F. Legrand, V. Morel, L. Piouffre, J. Bodmer, W.F. Bodmer, B. Bonne-Tamir, A. Cambon-Thomsen, Z. Chen, J. Chu, C. Carcassi, L. Contu, R. Du, L. Excoffier, G.B. Ferrara, J.S. Friedlander, H. Groot, D. Gurwitz, T. Jenkins, R.J. Herrera, X. Huang, J. Kidd, K.K. Kidd, A. Langaney, A.A. Lin, S.Q. Mehdi, P. Parham, A. Piazza, M.P. Pistillo, Y. Qian, Q. Shu, J. Xu, S. Zhu, J.L. Weber, H.T. Greely, M.W. Feldman, G. Thomas, J. Dausset, and L.L. Cavalli-Sforza. 2002. A human genome diversity cell line panel. *Science* 296:261–62.

Cavalli-Sforza, L.L. 1971. Use of models: Similarities and dissimilarities of sociocultural and biological evolution. (See above, ch. 4.)

——. 1992. Forty years ago in genetics: The unorthodox mating behavior of bacteria. *Genetics* 132:635–37

——. 1992. La matematica nella ricerca e nell'insegnamento biologici. I modelli matematici in biologia. *Bolletino U.M.I.* 7:313–33.

Cavalli-Sforza, L.L., A.C. Wilson, C.R. Cantor, R.M. Cook-Deegan, and M. C. King. 1991. Call for a worldwide survey of human genetic diversity. (See above, ch. 7.)

Cavalli-Sforza, L. L., A. Moroni, and G. Zei. 2004. *Consanguinity, Inbreeding, and Genetic Drift in Italy*. Princeton: Princeton University Press.

Diamond, J. and P. Bellwood. 2003. Farmers and their languages: The first expansions. *Science* 300:597–603.

Du, R., C. Xiao, and L.L. Cavalli-Sforza. 1997. Genetic distances between Chinese populations calculated on gene frequencies of 38 loci. *Science in China (Series C)* 40:613–21.

Edmonds, C. A., A. S. Lillie, and L. L. Cavalli-Sforza. 2004. Mutations arising in the wave front of an expanding population. *Proceedings of the National Academy of Sciences (USA)* 101:975–79.

Hewlett, B.S., A. de Silvestri, and C.R. Guglielmino. 2002. Semes and genes in Africa. (See above, ch. 4.)

Ke, Y., B. Su, X. Song, D. Lu, L. Chen, H. Li, C. Qi, S. Marzuki, R. Deka, P. Underhill, C. Xiao, M. Shriver, J. Lell, D. Wallace, R. S. Wells, M. Seielstad, P. Oefner, D. Zhu, J. Jin, W. Huang, R. Chakraborty, Z. Chen, and L. Jin. 2001. African origin of modern humans in East Asia: A tale of 12,000 Y chromosomes. *Science* 292:1151–53.

King, R. and P. A. Underhill. 2002. Congruent distribution of Neolithic painted pottery and ceramic figurines with Y-chromosome lineages. *Antiquity* 76:707–714.

Klein, J. and N. Takahata. 2002. *Where Do We Come From?* Berlin: Springer-Verlag.

Risch, N. (and 30 others). 1999. A genomic screen of autism: Evidence for a multilocus etiology. *American Journal of Human Genetics* 65:493–507.

Rosenberg, N. A., J. K. Pritchard, J. L. Weber, H. M. Cann, K. K. Kidd, L. A. Zhivotovsky, and M. W. Feldman. 2002. Genetic structure of human populations. *Science* 298: 2381–85.

Stent, G. S. and R. Calendar. 1971. *Molecular Genetics: An Introductory Narrative.* Berkeley: University of California Press.

Takahata, N., S-H Lee, and Y. Satta. 2001. Testing multiregionality of modern human origins. *Molecular Biology and Evolution* 18:172–83.

White, T. D., B. Asfaw, D. DeGusta, H. Gilbert, G. D. Richards, G. Suwa, and F. C. Howell. 2003. Pleistocene *Homo sapiens* from Middle Awash, Ethiopia. *Nature* 423:742–47.

Index

Note: Numbers in italics indicate an illustration.